W9-BJZ-584
COMPUTING

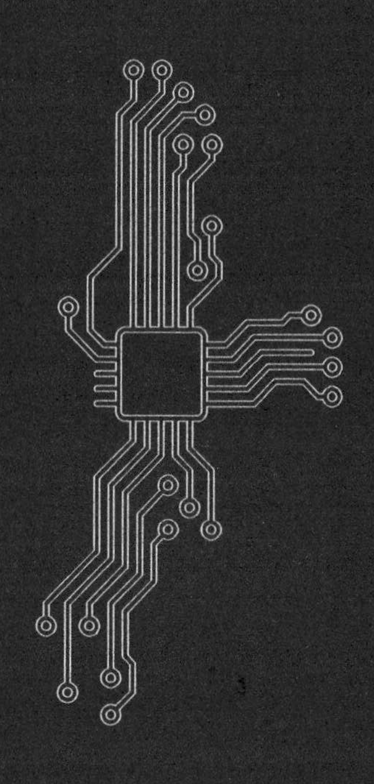

The MIT Press Essential Knowledge Series

Computing: A Concise History, Paul E. Ceruzzi
Information and the Modern Corporation, James Cortada
Intellectual Property Strategy, John Palfrey
Open Access, Peter Suber

This book was set in Chaparral Pro by the MIT Press. Printed and bound in the United States of America.

Library of Congress Cataloging-in-Publication Data

Cerruzi, Paul E.
Computing : a concise history / Paul E. Ceruzzi.
p. cm. —(MIT Press essential knowledge)
Includes bibliographical references and index.
ISBN 978-0-262-51767-6 (pbk. : alk. paper) 1. Computer science—History. I. Title.
QA76.17.C467 2012
004—dc23 2011053181

10

COMPUTING

A CONCISE HISTORY

PAUL E. CERUZZI

The MIT Press | Cambridge, Massachusetts | London, England

CONTENTS

SERIES FOREWORD

The MIT Press Essential Knowledge series presents short, accessible books on need-to-know subjects in a variety of fields. Written by leading thinkers, Essential Knowledge volumes deliver concise, expert overviews of topics ranging from the cultural and historical to the scientific and technical. In our information age, opinion, rationalization, and superficial descriptions are readily available. Much harder to come by are the principled understanding and foundational knowledge needed to inform our opinions and decisions. This series of beautifully produced, pocket-sized, soft-cover books provides in-depth, authoritative material on topics of current interest in a form accessible to nonexperts. Instead of condensed versions of specialist texts, these books synthesize anew important subjects for a knowledgeable audience. For those who seek to enter a subject via its fundamentals, Essential Knowledge volumes deliver the understanding and insight needed to navigate a complex world.

Bruce Tidor
Professor of Biological Engineering and Computer Science
Massachusetts Institute of Technology

INTRODUCTION

A familiar version of Zeno's paradox states that it is impossible for a runner to finish a race. First, he must traverse one-half of distance to the finish, which takes a finite time; then he must traverse one-half of the remaining distance, which takes a shorter but also finite time; and so on. To reach the finish line would thus require an infinite number of finite times, and so the race can never be won. The history of computing likewise can never be written. New developments transform the field while one is writing, thus rendering obsolete any attempt to construct a coherent narrative. A decade ago, historical narratives focused on computer hardware and software, with an emphasis on the IBM Corporation and its rivals, including Microsoft.[1] That no longer seems so significant, although these topics remain important. Five years ago, narratives focused on "the Internet," especially in combination with the World Wide Web and online databases. Stand-alone computers were important, but the network and its effects were the primary topics of interest. That has changed once again, to an emphasis on a distributed network of handheld devices, linked to a cloud of large databases, video, audio, satellite-based positioning systems, and more. In the United States the portable devices are called "smart phones"—the name coming from the devices from which they descended—but making phone calls seems to be the least interesting thing they do. The emphasis on IBM, Microsoft, and Netscape has given way to narratives that place Google and Apple at the center. Narratives are compelled to mention Facebook and Twitter at least once in every paragraph. Meanwhile the older technologies, including mainframe computers, continue to hum along in the background. And the hardware on which all of this takes place continues to rely on a device, the microprocessor, invented in the early 1970s.

Mathematicians have refuted Zeno's paradox. This narrative will also attempt to refute Zeno's paradox as it tells the story of the invention and subsequent development of digital technologies. It is impossible to guess what the next phase of computing will be, but it is likely that whatever it is, it will manifest four major threads that run through the story.

The Digital Paradigm

The first of these is a digital paradigm: the notion of coding information, computation, and control in binary form, that is, a number system that uses only two symbols, 1 and 0, instead of the more familiar decimal system that human beings, with their ten fingers, have used for millennia. It is not just the use of binary arithmetic, but also the use of binary logic to control machinery and encode instructions for devices, and of binary codes to transmit information. This insight may be traced at least as far back as George Boole, who described laws of logic in 1854, or before that to Gottfried Wilhelm Leibinz (1646–1716). The history that follows discusses the often-cited observation that "digital" methods of calculation prevailed over the "analog" method. In fact, both terms came into use only in the 1930s, and they never were that distinct in that formative period. The distinction is valid and is worth a detailed look, not only at its origins but also how that distinction has evolved.

Convergence

A second thread is the notion that computing represents a convergence of many different streams of techniques, devices, and machines, each coming from its own separate historical avenue of development. The most recent example of this convergence is found in the smart phone, a merging of many technologies: telephone, radio, television, phonograph, camera, teletype, computer, and a few more. The computer, in turn, represents a convergence of other technologies: devices that calculate, store information, and embody a degree of automatic control. The result, held together by the common glue of the digital paradigm, yields far more than the sum of the individual

The emphasis on IBM, Microsoft, and Netscape has given way to narratives that place Google and Apple at the center. Narratives are compelled to mention Facebook and Twitter at least once in every paragraph. Meanwhile, older technologies, including mainframe computers, continue to hum along in the background. And the hardware on which all of this takes place continues to rely on a device—the microprocessor—invented in the early 1970s.

parts. This explains why such devices prevail so rapidly once they pass a certain technical threshold, for example, why digital cameras, almost overnight around 2005, drove chemical-based film cameras into a small niche.

Solid-State Electronics

The third has been hinted at in relation to the second: this history has been driven by a steady advance of underlying electronics technology. That advance has been going on since the beginning of the twentieth century; it accelerated dramatically with the advent of solid-state electronics after 1960. The shorthand description of the phenomenon is "Moore's law": an empirical observation made in 1965 by Gordon Moore, a chemist working in what later became known as Silicon Valley in California. Moore observed that the storage capacity of computer memory chips was increasing at a steady rate, doubling every eighteen months. It has held steady over the following decades. Moore was describing only one type of electrical circuit, but variants of the law are found throughout this field: in the increase in processing speeds of a computer, the capacity of communications lines, memory capacities of disks, and so forth. He was making an empirical observation; the law may not continue, but as long as it does, it raises interesting questions for historians. Is it an example of "technological determinism": that technological advances drive history? The cornucopia of digital devices that prevails in the world today suggests it is. The concept that technology drives history is anathema to historians, who argue that innovation also works the other way around: social and political forces drive inventions, which in turn shape society. The historical record suggests that both are correct, a paradox as puzzling as Zeno's but even harder to disentangle.

The Human-Machine Interface

The final thread of this narrative concerns the way human beings interact with digital devices. The current jargon calls this the user interface. It goes to the philosophical roots of computing and is one reason that the subject is so fascinating. Are we trying to create a mechanical replacement for a human being, or a tool that works in symbiosis with humans, an extension of the human's mental faculties? These debates, once found only among science-fiction writers, are not going away and will only increase as computers become more capable, especially as they acquire an ability to converse in natural language. This theme is a broad one: it ranges from philosophical implications about humanity to detailed questions about machine design. How does one design a device that humans can use effectively, that takes advantage of our motor skills (e.g., using a mouse or touch screen) and our ability to sense patterns (e.g., icons), while providing us with information that we are not good at retaining (e.g., Google or Wikipedia)? We shall see that these detailed questions of human use were first studied intensively during World War II, when the problem of building devices to compute and intercept the paths of enemy aircraft became urgent.

With those four themes in mind—digitization, convergence, solid-state electronics, and the human interface—what follows is a summary of the development of the digital information age.

1

THE DIGITAL AGE

In the spring of 1942, as World War II was raging, the U.S. National Defense Research Committee convened a meeting of scientists and engineers to consider devices to aim and fire anti-aircraft guns. The *Blitzkrieg*, a brilliant military tactic based on rapid attacks by German dive bombers, made the matter urgent. The committee examined a number of designs, which they noticed fell into two broad categories. One directed antiaircraft fire by constructing a mechanical or electrical analog of the mathematical equations of fire control, for example, by machining a camshaft whose profile followed an equation of motion. The other solved the equations numerically—as with an ordinary calculating machine, only with fast electrical pulses instead of mechanical counters. One member of the committee, Bell Telephone Laboratories mathematician George Stibitz, felt that the term *pulse* was not quite right. He suggested another term that he felt was more descriptive: *digital*. The word referred to the method of counting on one's fingers, or digits. It has since become the adjective that defines social, economic, and political life in the twenty-first century.[1]

It took more than just the coining of a term to create the digital age, but that age does have its origins in secret projects initiated or conducted during World War II. Most histories of computing, which purport to cover the full range of the topic, do not explain how such an invention, intended as a high-speed replacement for calculators during the war, could have led to such a far-reaching social impact.

Nor do those wartime developments, as significant as they were, explain the adoption of digital techniques for communications. That took place not during World War II but two decades later, when an agency of the U.S. Defense Department initiated a program to interconnect defense computers across the United States. That combination of computing and communications unleashed a flood of social change, in the midst of which we currently live.

Telecommunications, like computing, has a long and well-documented history, beginning with the Morse and Wheatstone electric telegraphs of the mid-nineteenth century, followed by the invention and spread of the telephone by Alexander Graham Bell, Elisha Gray, Thomas Edison, and others later that century. What was different about the 1960s computer networks? Nearly every social and business phenomenon we associate with the Internet was anticipated by similar uses of the telegraph a century earlier.[2] The telegraph, combined with the undersea cable, did transform society, yet the transformation effected by the more recent application of a digital paradigm seems to be many times greater.

It is dangerous to apply modern terms to events of the past, but one may violate this rule briefly to note that the electric telegraph, as refined by Samuel Morse in the 1840s, was a proto "digital" device. It used pulses, not continuous current, and it employed a code that allowed it to send messages rapidly and accurately over long distances with a minimum number of wires or apparatus. Typesetters had long known that certain letters (e.g., *e, t, a*) were used more frequently than others, and on that basis the codes chosen for those letters were shorter than the others. A century later mathematicians placed this ad hoc understanding of telecommunications on a theoretical basis. What came to be called information theory emerged at the same time as, and independent of, the first digital computers in the 1930s and 1940s. Binary (base-2) arithmetic, bits, and bytes (8-bit coded characters) are familiar at least in name to modern users of digital devices. The roots of that theory are less well known, but those roots made the modern digital age possible.

Many histories of computing begin with Charles Babbage, the Englishman who tried, and failed, to build an

"Analytical Engine" in the mid-nineteenth century—the same time as Morse was developing the telegraph.[3] The reason is that Babbage's design—what we now call the architecture of his machine—was remarkably modern. It embodied the basic elements that were finally realized in the computers built during World War II. We now see, however, that to begin with Babbage is to make certain assumptions. What exactly is a "computer"? And what is its relation to the digital age that we are living in today?

The Components of Computing

Computing represents a convergence of operations that had been mechanized to varying degrees in the past. Mechanical aids to calculation are found in antiquity, when cultures developed aids to counting and figuring such as pebbles (Latin *calculi*, from which comes the term *calculate*), counting boards (from which comes the modern term *countertop*), and the abacus—all of which survive into this century. Although it may seem arbitrary, the true mechanization of calculation began when inventors devised ways not only to record numbers but to add them, in particular to automatically carry a digit form one column to the next when necessary, especially for carries like 999 + 1. That began with Pascal's adding machine of 1642, or with a device invented by Wilhelm Schickard in 1623. Leibniz extended

Pascal's invention by developing a machine, a few decades later, that could multiply as well as add. The mechanisms by which these devices operated lay dormant until the nineteenth century, when advancing commerce and business created a demand that commercial manufacturers sought to fill. Toward the end of that century, mechanical calculators of intricate design appeared in Europe and in the United States. The Felt Comptometer, invented in the 1880s, was one of the first successful calculators, owing to its simplicity, speed, and reliable operation. The Burroughs adding machine, invented by William S. Burroughs around the same time, also was a commercial success. Burroughs survived as a supplier of electronic computers into the 1980s and is the ancestor of today's Unisys. In Europe, machines supplied by companies including Brunsviga and Odhner were also widely sold. On these machines, the user set numbers on a set of wheels rather than press keys, but they worked on similar principles.

As significant as calculation were two additional functions: the automatic storage and retrieval of information in coded form and the automatic execution of a sequence of operations. That is the reason historians began with the Analytical Engine that Charles Babbage attempted to build in the nineteenth century. Babbage never completed that machine, for reasons that only in part had to do with the state of mechanical engineering at the time. In the 1830s, when Babbage was sketching out ideas for such an engine, neither he nor anyone else could draw on electrical technology to implement his ideas; everything had to be done mechanically. Given the necessary level of complexity that a computer must have, a mechanical computer of decent power was not practical then, and not today either. The recent successful reconstruction, at great expense, of Babbage's other project, the Difference Engine, proves this point.[4] It works, but the Difference Engine's capabilities are nowhere near those of the Analytical Engine. An analogy might be to compare Babbage's vision with Leonardo's sketches for a flying machine: Leonardo's vision was sound, but heavier-than-air flight had to await the invention of the gasoline engine to provide sufficient power in a lightweight package.

By this measure, one might begin the history of computing in the late nineteenth century, when the American inventor Herman Hollerith developed, for the 1890 U.S. Census, a method of storing information coded as holes punched into cards. Hollerith developed not just the punched card but a suite of machines that used cards to sort, retrieve, count, and perform simple calculations on data punched onto cards. The devices he developed combined complex mechanisms with electromagnets and motors to perform operations. The use of electricity was not required. A rival inventor, James Powers, produced pure mechanical punched card equipment to avoid infringing on Hollerith's parents, but in practice the flexibility that

Hollerith's use of electricity gave his machines was an advantage as the machinery was called on to perform ever more complex operations. By the time of World War II, electrical circuits took on an even greater significance, not just as a carrier of information but also as a way of performing computing operations at high speeds—a property that in theory is not required of a true computer but in practice is paramount.

The inherent flexibility of Hollerith's system of machines built around the punched card led to many applications beyond that of the U.S. Census. Hollerith founded the Tabulating Machine Company to market his inventions; it was later combined with other companies to form the Computing-Tabulating-Recording Company (C-T-R), and in 1924, the new head of C-T-R, Thomas Watson, changed the name to the International Business Machines Corporation, today's IBM. In 1927 the Remington Rand Corporation acquired the rival Powers Accounting Machine Company, and these two would dominate business accounting for the next four decades.

It is not known where Hollerith got the idea of storing information in the form of holes punched onto card stock, but the concept was not original with him. Babbage proposed using punched cards to control his Analytical Engine, an idea he borrowed from the looms invented by the Frenchman Joseph-Marie Jacquard (1752–1834), who in the nineteenth century used punched cards to control the weaving of cloth by selectively lifting threads according to a predetermined pattern (Jacquard cloth is still woven to this day). Jacquard looms were common in Hollerith's day, so he was probably familiar with the way punched cards controlled them. However, there is a crucial difference between Jacquard's and Hollerith's systems: Jacquard used cards for control, whereas Hollerith used them for storage of data. Eventually IBM's punched card installations would also use the cards for control. It is fundamental to the digital paradigm that information stored in digital form can be used for storage, control, or calculation, but an understanding of that would not come until decades later. Before World War II, the control function of a punched card installation was carried out by people: they carried decks of cards from one device to another, setting switches or plugging wires on the devices to perform specific calculations, and then collecting the results.

The concept of automatic control, the ancestor of what we now call software, is a third component of computing, and it too has a history that can be traced back to antiquity. Jacquard's invention was an inside-out version of a device that had been used to control machinery for centuries: a cylinder on which were mounted pegs, which tripped levers as it rotated. These had been used in medieval clocks that executed complex movements at the sounding of each hour; they are also found in wind-up toys, including music boxes. Babbage's Analytical Engine was to contain a number of such cylinders to carry more detailed sequences of operations as directed by the punched cards; today we might call this the computer's microprogramming, or read-only memory (ROM). Continuous control of many machines, including classic automobile engines, is effected by cams, which direct the movement of other parts of the machine in a precisely determined way. Unlike cylinders or camshafts, punched cards can be stacked in an arbitrarily long sequence. It is also easy to substitute a short sequence of cards in the stack to tailor the machine for a specific problem, but Jacquard looms used cards that were tied to one another, making any modification to the control difficult.

Control, storage, calculation, the use of electrical or electronic circuits: these attributes, when combined, make a computer. To them we add one more: communication—the transfer of coded information by electrical or electronic means across geographical distances. This fifth attribute was lacking in the early electronic computers built in the 1930s and 1940s. It was the Defense Department's Advanced Research Projects Agency (ARPA)'s mission, beginning in the 1960s, to reorient the digital computer to be a device that was inherently networked, for which communication was as important to it as calculation, storage, or control.

The origins of the electric telegraph and telephone are well known, but their relationship to computing is complex. In 1876, Alexander Graham Bell publicly demonstrated a telephone: a device that transmitted the human voice over wires. The telephone's relationship to the invention of the computer was indirect. Computers today operate by electrical circuits that allow only one of two states: in modern terms they are both digital, as described above, and binary: they count in base 2. The telephone operated by inducing a continuous variation of current based on the variations of the sound of a person's voice: in today's terms, it was an analog device. Like *digital*, that term was also unknown before the late 1930s, and therefore not entirely proper to use here. Devices that compute by analogy were once common. The slide rule, for example, was in common use into the 1970s, when it was driven out by the pocket calculator. During the first decades of electronic digital computing, the 1940s and 1950s, there were debates over the two approaches, with analog devices fading into obscurity. Nature is in a fundamental sense continuous, as in the infinite variations of the human voice or the sounds of a musical instrument. But the digital paradigm has prevailed, even in telephony. During World War II, Bell Telephone Laboratories developed a machine that translated voice signals into discrete pulses, encoded them, and reconstituted the voice at the other end—this was to enable Franklin D. Roosevelt and Winston Churchill to speak to each other securely.[5] That was a one-time development, although eventually all phone calls were encoded this way, in what is now called pulse code modulation. The technique is used not so much for secrecy (although that can be done when necessary), but to exploit of the inherent advantages of digital electronics.

Bell's successful defense of his patent, and the subsequent establishment of a wealthy, regulated monopoly to provide telephone service in the United States, led to generous funding for the Bell Telephone Laboratories, which conducted fundamental research in the transmission of information, broadly defined. The role of one Bell Labs mathematician, George Stibitz, has already been mentioned. It was a team of Bell Labs researchers who invented the transistor in the 1940s; two decades later, another Bell Labs team developed the Unix operating system, to mention only the most visible fruits of Bell Laboratories' research. And it was at Bell Labs where much of the theory of information coding, transmission, and storage was developed.

Once again: the modern computer is a convergence of separate streams of information handling, each with its own rich tradition of technological history. Each of the streams described thus far played a major role. One could add other antecedents such as the development of radio, motion pictures, and photography. The line of mechanical calculators seems to be at the forefront, yet it was the Hollerith system of punched cards, centered around a machine called the tabulator, that had a greater influence. Besides the tabulator, two other critical devices were employed: a key punch, by which a human operator keyed in data, and a sorter, which sorted cards based on the presence or absence of a hole in a desired column. In the early decades of the twentieth century, other devices were added to a typical installation, but these were the main ones.

From Tabulator to Computer, 1890–1945

The tabulator kept a record of how many cards had a hole punched in each of its columns. The early tabulators recorded the numbers on a dial that resembled a clock face; later a more familiar counter was used. Although it is hard to imagine that such a basic function could be so important, it was not until the 1920s that other arithmetic functions were provided. What made the tabulator so important was the flexible way it could be used, based on the information punched into different columns, and the way its use could be combined with the other devices, especially the sorter, to perform what we would now call sophisticated data processing. Information, once it was punched onto a card, could be used and reused in many ways. That roomful of equipment was what the early electronic computers replicated; its "program" was carried out by human beings carrying decks of cards from one machine to another and changing the settings on the various machines as the cards were run through them.

Communications, within and outside that room, was also present, in an ad hoc fashion. Human beings carried data from one machine to another as decks of cards. The electric telegraph carried information to and from the installation. Among the first, outside the government, to adopt punched card accounting in the United States were railroads. And railroads were early adopters of the telegraph as well, because they were the first significant business whose managers had to coordinate operations over wide geographical areas. The railroad rights of way became a natural corridor for the erection of wires across the continent, to an extent that people assume the two technologies could not have existed without each other. That is an exaggeration but not far from the truth. Although times have changed, modern overland Internet traffic is carried on fiber-optic lines, which are often laid underground (not on poles) along railroad rights of way.

If the telegraph apparatus did not physically have a presence in the punched card installation, its information was incorporated into the data processed in that room. Railroad operators were proficient at using the Morse code and were proud of their ability to send and receive the dots and dashes accurately and quickly. Their counterparts in early commercial and military aviation did the same, using the "wireless" telegraph, as radio was called. What worked for railroads and aircraft was less satisfactory for other businesses, however. Among the many inventions credited to Thomas Edison was a device that printed stock market data sent by telegraph on a "ticker tape," so named because of the sound it made. The physical ticker tape has been replaced by electronic displays, but the terse symbols for the stocks and the "crawling" data have been carried over into the electronic era.[6] Around 1914, Edward E. Kleinschmidt, a German immigrant to the United States, developed a series of machines that combined the keyboard and printing capabilities of a typewriter with the ability to transmit messages over wires.[7] In 1928 the company he founded changed its name to the Teletype Corporation, which AT&T purchased two years later. AT&T, the telephone monopoly, now became a supplier of equipment that transmitted text as well as voice (see figure 1.1).

The Teletype (the name referred to the machine as well as to the company that manufactured it) was primitive by today's standards: slow, noisy, and with few symbols other than the uppercase letters of the alphabet and the digits 0 through 9. But it found a large market, providing the communications component to the information processing ensemble described above. The machine took its place alongside other information handling equipment in offices, the government, and the military. It also entered our culture. Radio newscasters liked to have a Teletype chattering in the background as they read the news, implying that what they were reading was "ripped from the wires." Jack Kerouac did not type the manuscript for his

Figure 1.1 Computing at Bell Laboratories designed by George Stibitz and using modified telephone switching equipment. (a) H. L. Marvin operating a Bell Labs specialized calculator using a modified Teletype, 1940. (b) Control panel for a Bell Labs computer used for fire control, circa 1950. (*Source*: Lucent/Alcatel Bell Laboratories)

Beat novel, *On the Road,* on a continuous reel of Teletype paper, but that is the legend. In the 1970s manufacturers of small computers modified the Teletype to provide an inexpensive terminal for their equipment. Bill Gates and Paul Allen, the founders of Microsoft, marketed their first software products on rolls of teletype tape. Among those few extra symbols on a Teletype keyboard was the @ sign, which in 1972 was adopted as the marker dividing a person's e-mail address from the computer system the person was using. Thus, we owe the symbol of the Internet age to the Teletype (see figure 1.2).

The Advent of Electronic Computing

In the worlds of U.S. commerce and government, these systems reached a pinnacle of sophistication in the 1930s, ironically at a time of economic depression. Besides IBM and Remington Rand supplying punched card equipment, companies like National Cash Register (later NCR), Burroughs, Victor Adding Machine (later Victor Comptometer), and others supplied mechanical equipment: adding machines, cash registers, accounting machines, time clocks, "computing" scales (which priced items by weight), duplicating machines, and electric typewriters. Supplementing these were systems that used little mechanization but were crucial to controlling the flow of information: card filing systems, standardized forms in multiple copies with carbon paper, bound or loose-leaf ledgers, and more.[8] The digital computer upended this world, but it did not happen overnight. Many of the firms noted here entered the commercial computer industry, with IBM, Remington Rand, and Burroughs among the top U.S. computer manufacturers through the 1960s.[9]

As is so often the case in the history of technology, at the precise moment when this system was functioning at its peak of efficiency, digital computing emerged to replace it. The most intense period of innovation was during World War II, but as early as the mid-1930s, the first indications of a shift were apparent. In hindsight, the reasons were clear. The systems in place by the 1930s were out of balance. The versatility of the punched card, which stored data that could be used and reused in a variety of ways, required that human beings first of all make a plan for the work that was to be done and then carry out that plan by operating the machines at a detailed level. Human beings had to serve as the interface with the adding machines, accounting machines, and other equipment, and with the nonmechanized patterns of paper flow that corporations and government agencies had established. For example, an eighty-column card, in spite of its great flexibility, was ill suited to store or print a person's mailing address. Thus, a company that periodically sent out bills required another machine, such as the American Addressograph: a metal plate on which the person's name and

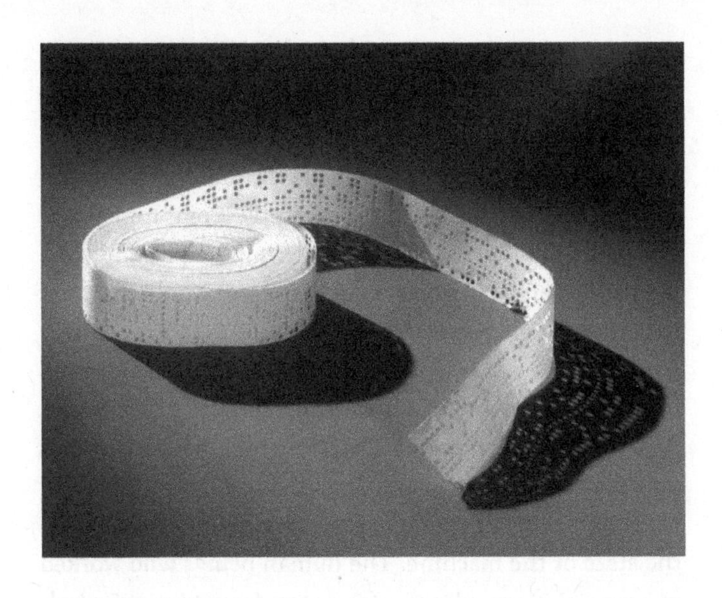

Figure 1.2 (left) The Teletype ASR-33. The Teletype had only uppercase letters, numbers, and a few special characters. In 1972 Ray Tomlinson, an engineer at Bolt Beranek and Newman in Cambridge, Massachusetts, chose the @ sign (shift-p) to separate the recipient of an e-mail message from the host machine to which that message was addressed; it has since become the symbol of the Internet age. (*Source*: Digital Equipment Corporation, now Hewlett-Packard) (above) Teletypes were used as the main input-output device for the early personal computers, until inexpensive video terminals became available. This piece of Teletype tape contains an interpreter for the BASIC programming language, Microsoft's first product. (Credit: Smithsonian Institution)

address were embossed, from which mailing labels were printed. A human being had to coordinate the operation of these two technologies.[10]

For many problems, especially those in science or engineering, a person operating a simple Comptometer or calculator could perform arithmetic quite rapidly, but she (and such persons typically were women) would be asked to carry out one sequence if interim results were positive, a different one if negative. The plan for her work would be specified in detail in advance and given to her on paper. Punched card equipment also had stops built into their operation, signaling to the operator to remove a deck of cards and proceed in a different direction depending on the state of the machine. The human beings who worked in some of these places—for example, astronomical observatories where data from telescope observations were reduced—had the job title "computer": a definition that was listed as late as the 1970 edition of *Webster's New World Dictionary*. From the human computers, who alone had the ability to direct a sequence of operations as they worked, the term *computer* came to describe a machine.

A second reason that the systems of the 1930s fell short emerged during World War II: a need for high speed. The systems of the 1930s used electricity to route signals and transfer information from one part of a device to another and to communicate over long distances. Communication over the telegraph proceeded at high speed, but calculations were done at mechanical speeds, limited by Newton's laws that relate speed to the amount of energy required to move a piece of metal. As the complexity of problems increased, the need for high-speed operation came to the fore, especially in wartime applications like breaking an enemy code or computing the trajectory of a shell. High speeds could be achieved only by modifying another early twentieth-century invention, the vacuum tube, which had been developed for radio and telephone applications. Those modifications would not be easy, and substituting tubes for mechanical parts introduced a host of new problems, but with electronics, the calculations would not be bound not by Newton's laws, thus allowing calculating as well as computing to approach the speed of light.

These limiting factors, of automatic control and of electronic speeds, were "reverse salients," in Thomas Hughes's term: impediments that prevented the smooth advance of information handling on a broad front (the term comes from World War I military strategy).[11] Solutions emerged simultaneously in several places beginning around 1936. The work continued during World War II at a faster pace, although under a curtain of secrecy, which in some respects mitigated the advantages of having funds and human resources made available to the computer's inventors.

2

THE FIRST COMPUTERS, 1935–1945

In summer 1937, Konrad Zuse was a twenty-seven-year-old mechanical engineer working at the Henschel Aircraft Company in Berlin. Under the Nazi regime, Germany was arming itself rapidly, although Zuse recalled that neither he nor his fellow young engineers foresaw the war and destruction that would come two years later. He was occupied with tedious calculations relating to the design of aircraft. He began work, on his own time, on a mechanical calculator that would automate that process. Zuse was one of several engineers, scientists, and astronomers across Europe and in the United States who were thinking along the same lines as they struggled with the limitations of existing calculating devices. In June 1937, however, he made a remarkable entry in his diary: "For about a year now I have been considering the concept of a mechanical brain.... Discovery that there are elementary operations for which all arithmetic and thought processes can be solved.... For every problem to be solved there must be a special purpose brain that solves it as fast as possible."[1]

Zuse was a mechanical engineer. He chose to use the binary, or base-2, system of arithmetic for his proposed calculator, because as an engineer, he recognized the inherent advantages of switches or levers that could assume one of only two, instead of ten, positions. But as he began sketching out a design, he had an insight that is fundamental to the digital age that has followed: he recognized that the operations of calculation, storage, control, and transmission of information, until that time traveling on separate avenues of development, were in fact one and the same. In particular, the control function, which had not been mechanized as much as the others in 1937, could be reduced to a matter of (binary) arithmetic. That was the basis for his use of the terms *mechanical brain* and *thought processes*, which must have sounded outrageous at the time. These terms still raise eyebrows when used today, but with each new advance in digital technology, they seem less and less unusual. Zuse realized that he could design mechanical devices that could be flexibly rearranged to solve a wide variety of problems: some requiring more calculation and others requiring more storage, though each requiring varying degrees of automatic control. In short, he conceived of a universal machine. Today we are familiar with its latest incarnation: a handheld device that, thanks to numerous third-party application programs called "apps," can do almost anything: calculate, play games, view movies, take photographs, locate one's position on a map, record and play music, send and process text—and by the way, make phone calls (see figure 2.1).

Zuse recalled mentioning his discovery to one of his former mathematics professors, only to be told that the theory Zuse claimed to have discovered had already been worked out by the famous Göttingen mathematician David Hilbert and his students.[2] But that was only partially true: Hilbert had worked out a relationship between arithmetic and binary logic, but he did not extend that theory to a design of a computing machine. The Englishman Alan M. Turing (1912–1954) had done just that, in a thirty-six-page paper published in the *Proceedings of the London Mathematical Society* the year before, in 1936.[3] (Zuse was unaware of Turing's paper until years later; he learned of Babbage only when he applied for a German patent, and the patent examiner told him about Babbage's prior work.) So although Zuse took the bold step of introducing theoretical mathematics into the design of a mechanical calculator, Turing took the opposite but equally bold step: introducing the concept of a "machine" into the pages of a theoretical mathematics journal.

In his paper, Turing described a theoretical machine to help solve a problem that Hilbert himself had proposed at the turn of the twentieth century.[4] Solving one of those problems placed Turing among the elite of mathematicians.

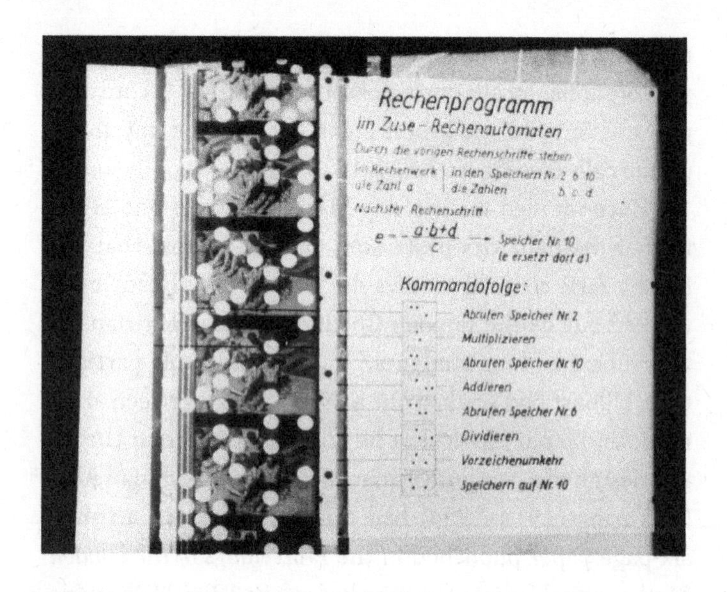

Figure 2.1 Programs for a computer built in Berlin by Konrad Zuse, circa 1944, using discarded movie film. Zuse's program-controlled calculators may have been the first to realize Babbage's vision of an automatic calculator. Most of Zuse's work was destroyed during World War II, but one of his machines survived and was used into the 1950s. (Credit: Konrad Zuse)

Mathematicians admired his solution, but it was his construction of this "machine" that placed Turing among the founders of the digital age. I put the term in quotation marks because Turing built no hardware; he described a hypothetical device in his paper. It is possible to simulate the workings of a Turing machine on a modern computer, but in only a restricted sense: Turing's machine had a memory of infinite capacity, which Turing described as a tape of arbitrary length on which symbols could be written, erased, or read. No matter. The machine he described, and the method by which it was instructed to solve a problem, was the first theoretical description of the fundamental quality of computers. That is, a computer can be programmed to perform an almost infinite range of operations if human beings can devise formal methods of describing those operations. The classical definition of *machine* is of a device that does one thing and one thing well; a computer is by contrast a universal machine, whose applications continue to surprise its creators. Turing formalized what Zuse had recognized from an engineer's point of view: a general-purpose computer, when loaded with a suitable program, becomes "a special purpose brain" in Zuse's words that does one thing—whatever the programmer wants it to do.[5]

Beginning in the mid-1930s, as people began designing machines that could carry out a sequence of operations, they rediscovered Babbage's prior work. In fact it had never been lost. Descriptions of Babbage's designs had been published and were available in research libraries, and fragments of his machines had been preserved in museums. But because Babbage had failed to complete a general-purpose computer, many concluded that the idea itself was flawed. Babbage had not gotten far enough in his work to glimpse the theories that Turing, and later John von Neumann, would develop. He did, however, anticipate the notion of the universality of a programmable machine. In a memoir published in 1864 (for many years out of print and hard to find), Babbage remarked: "Thus it appears that the whole of conditions which enable a *finite* machine to make calculations of *unlimited* extent are fulfilled in the Analytical Engine. . . . I have converted the infinity of space, which was required by the conditions of the problem, into the infinity of time."[6]

A contemporary of Babbage, Ada Augusta, had the same insight in annotations she wrote for an Italian description of Babbage's engine. On that basis Augusta is sometimes called the world's "first" programmer, but that gives her too much credit. Nevertheless, she deserves credit for recognizing that a general-purpose programmable calculator by nature is nothing like the special-purpose devices that people associated with the term *machine.* One can only speculate how she might have elaborated on this principle had Babbage gotten further with his work.

There was a renewed interest in such machines in the mid-1930s. But then it took another twenty years before the hardware could get to a level of operations where these theoretical properties could matter. Until about 1950, it was a major accomplishment if one could get an electronic computer to operate without error for even a few hours. Nonetheless, Turing's insight was significant.[7] By the late 1940s, after the first machines based on ad hoc designs began working, there was a vigorous debate about computer design among engineers and mathematicians. From those debates emerged a concept, known as the stored program principle, that extended Turing's ideas into the design of practical machinery. The concept is usually credited to the Hungarian mathematician John von Neumann (1903–1957), but von Neumann's description of it came only after a close collaboration with American engineers J. Presper Eckert and John Mauchly at the University of Pennsylvania. And von Neumann was familiar with Turing's work at least as early as 1938.[8] Modern computers store both their instructions—the programs—and the data on which those instructions operate in the same physical memory device, with no physical or design barrier between them. Computers do so for practical reasons: each application may require a different allocation of memory for each, so partitioning the storage beforehand is unwise. And computers do so for theoretical reasons: programs and data are treated the same inside the machinery because fundamentally programs and data *are* the same.

Zuse eventually built several computing devices, which used a mix of mechanical and electromechanical elements (e.g., telephone relays) for calculation and storage. He used discarded movie film, punched with holes, to code a sequence of operations. His third machine, the Z3 completed in 1941, was the first to realize Babbage's dream. Its existence was little known outside Germany until well after the war's end, but Zuse eventually was given credit for his work, not only in building this machine but also in his theoretical understanding of what a computer ought to be. The mainstream of computing development, however, now moved to the United States, although the British had a head start. The early American computers were designed with little theoretical understanding. The notion of basing a design on the principles of symbolic logic, whether from Boole, Hilbert, or others, was not adopted until the early 1950s, when California aerospace companies began building digital computers for their specialized aerospace needs.

Along with Zuse, other mathematicians, astronomers, and engineers began similar efforts in the late 1930s. Some adopted punched cards or perforated paper tape—modifications of existing Teletype or IBM storage media—to encode a sequence of operations, which a corresponding control mechanism would read and transmit to a calculating mechanism.[9] Many used IBM tabulators or telephone switching equipment, which transmitted signals electrically at high speed but calculated mechanically at a lower speed. Other projects used vacuum tubes to do the actual counting, which increased the overall speed of computation by several hundred-fold or more. It is from these latter experiments that the modern electronic digital computer has descended, but the early attempts to compute electronically must be considered in the context of the initial attack on the need to automate the procedure for solving a problem.

Of the many projects begun in that era, I mention only a few. I have selected these not so much because they represent a first of some kind, but because they illustrate the many different approaches to the problem. Out of these emerged a configuration that has survived into this century through all the advances in underlying technology.

In 1934, at Columbia University in New York City, Wallace Eckert founded a laboratory that used IBM equipment to perform calculations related to his astronomical research, including an extensive study of the motion of the moon, drawing on the earlier use of punched cards by the British astronomer L. J. Comrie. The two astronomers used the machines to do calculations that the machines were not designed for, taking advantage of the ability to store mathematical tables as decks of punched cards that could be used over and over without errors of transcription that might occur when consulting printed tables.[10] At first Eckert modified that equipment only slightly, but in the following decade, he worked with IBM to develop machinery that used cables and switches, controlled by cards that had commands, not data, punched on them. In essence, those cables and switches replicated the actions of the human beings who operated a 1930s-era punched card installation.[11]

While Eckert was at Columbia University, Howard Aiken, a physics instructor at Harvard University, faced a similar need for calculating machinery. He proposed a machine that computed sequences directly, as directed by a long strip of perforated paper tape. Aiken's machine was thus more in line with what Babbage had proposed, and indeed Aiken's proposal acknowledges Babbage's prior work. By the late 1930s, Aiken was able to take advantage of electrical switching technology, as well as the high degree of mechanical skills developed at IBM. Aiken's proposal, written in 1937, described in detail why he felt that existing punched card machines, even when modified, would not suit his needs.[12] Eckert had considered such a special-purpose calculator but resisted having one built, believing that such a device would be very expensive and would take several years to build. Eckert was correct: although Aiken's proposal came to fruition as the Automatic Sequence Controlled Calculator, unveiled publicly at Harvard in 1944, it did take several years to design and build (at IBM's Endicott, New York, laboratories). And it would not have been built had a world war not been raging and had Aiken not received financial support from the U.S. Navy.[13]

The Advent of Electronics

All of the machines described used electricity to carry signals, but none used electronic devices to do the actual calculation. Among those who first took that crucial step, around 1938, was J. V. Atanasoff, a professor of physics at Iowa State College in Ames. He came to that step from an investigation of ways to mechanize the solution of large systems of linear algebraic equations, which appear throughout many fields of physics and related sciences. The equations themselves were relatively simple—"linear" implying that they form a straight line when graphed. What was most intriguing about this problem was that a method of solution of the systems had been known and described mathematically for at least a century, and the method could be written as a straightforward sequence of operations. In other words, the procedure for their solution was an *algorithm*—a recipe that, if followed, guaranteed a solution. The solution of a large system required many such steps, however, and above a small threshold of complexity, it was impractical for human beings to carry them out. What was needed was, first of all, a way to perform arithmetic more rapidly than was done with mechanical calculators and, second, a method of executing the sequence of simple steps that would yield a solution.

It was to address the first of those problems that Atanasoff conceived of the idea of using vacuum tubes. Like his contemporary Zuse, he also saw the advantages of using the binary system of arithmetic, since it made the construction of the calculating circuits much easier. He did not carry that insight over to adopting binary logic for sequence control, though. Atanasoff's machine was designed to solve systems of linear equations, and it could not be programmed to do anything else. The proposed machine would have a fixed sequence coded into a rotating drum.[14]

In a proposal he wrote in 1940 to obtain support for the project, Atanasoff described a machine that computed at high speeds with vacuum tubes. He remarked that he considered "analogue" techniques but discarded them in favor of direct calculation (that is probably the origin of the modern term *analog* referring to computing machinery). With financial support from Iowa State College and engineering support from Clifford Berry, a colleague, he completed a prototype that worked, although erratically, by 1942. That year he left Iowa for the Washington, D.C., area, where he was pressed into service working on wartime problems for the navy. Thus, while the onset of World War II made available large sums of money and engineering resources for some computer pioneers, the war was a hindrance for others. Atanasoff never completed his machine. Had it been completed, it might have inaugurated the computer age a decade before that happened. Howard Aiken was also called away by the navy, but fortunately for him, IBM engineers and Harvard staff were well on their way to finishing his Sequence Controlled Calculator. In Berlin, Zuse was able to learn of the prior work of Babbage and study the international effort to develop mathematical logic, but after 1940, he worked in isolation, finding it difficult to get skilled workers and money to continue his efforts.

If the onset of war hindered Atanasoff's attempts to use vacuum tubes, it had the opposite effect in the United Kingdom, where at Bletchley Park, a country estate located in Buckinghamshire, northwest of London, multiple copies of a device called the "Colossus" were in operation by 1944. Details about the Colossus, indeed its very existence, remained a closely guarded secret into the 1970s. Some information about its operation and use remains classified, although one of the first tidbits of information to appear was that Alan Turing was involved with it. The work at Bletchley Park has an odd place in the history of computing. By the 1970s, several books on this topic appeared, and these books set a pattern for the historical research that followed. That pattern emphasized calculation: a lineage from Babbage, punched card equipment, and the automatic calculators built in the 1930s and 1940s. The

Colossus had little, if any, numerical calculating ability; it was a machine that processed text. Given the overwhelming dominance of text on computers and the Internet today, one would assume that the Colossus would be heralded more than it is. It has not, in part because it was unique among the 1940s-era computers in that it did no numerical calculation, which is implied by the definition of the word *computer*. By the time details about the Colossus became known, the notion of "first" computers had (rightly) fallen out of favor among historians. The Colossus did operate at electronic speeds, using vacuum tubes for both storage and processing of data. And its circuits were binary, with only two allowable states, and they exploited the ability to follow the rules of symbolic logic.[15]

The Colossus was not the only computing machine in use at Bletchley. Another machine, the Bombe, used mechanical wheels and electric circuits to decode German messages were encrypted by the Enigma machine. The Bombes were the result of an interesting collaboration between the mathematicians at Bletchley and engineers at the American firm National Cash Register in Dayton, Ohio, where much of the hardware was manufactured. The Enigma resembled a portable typewriter, and the Germans used it to scramble text by sending typed characters through a series of wheels before being transmitted. The Bombes were in a sense Enigma machines running in reverse, with sets of wheels that tested possible code combinations. The Colossus, by contrast, attacked messages that the Germans had encrypted electronically by a German version of the Teletype. One could say that the Bombes were reverse-engineered Enigmas, while the Colossi were protocomputers, programmed to decode teletype traffic.

The work at Bletchley shortened the war and may even have prevented a Nazi victory in Western Europe. The need for secrecy limited, but did not prevent, those who built and used the machines to transfer their knowledge to the commercial sector. The British did establish a computer industry, and many of those who worked at Bletchley were active in that industry, including pioneering work in marketing computers for business and commercial use. But the transfer of the technology to the commercial world was hindered, as the value of cryptography to national security did not end in 1945. It is as critical, and secret, today as it ever was.

The Colossus machines may have been destroyed at the end of the war, but even if they were not, there was no easy path to build a commercial version of them. The United States did a better job in transferring wartime computing technology to peaceful uses, although National Cash Register did not exploit its experience building the Bombes to further its business machines technology. After the war, American code breakers working for the navy used their experience to develop an electronic computer, later marketed and sold commercially by the Minneapolis firm Engineering Research Associates. The ERA 1101 was a general-purpose electronic computer, the marketing of which did not divulge secrets of how it may have been used behind security walls. In the United States, this work is concentrated at the National Security Agency (NSA), whose headquarters is at Fort Meade, Maryland. The ERA computers were not the only examples of technology transfer from the secret world of code breaking. In spite of its need for secrecy, the NSA has published descriptions of its early work in computing, enough to show that it was at the forefront of research in the early days.[16] We do not know where it stands today or how its work dovetails with, say, the parallel cutting-edge work on text analysis going on at places like Google, on the opposite coast of the United States.

Fire Control

At the same time that the development of machinery for code breaking was underway, an intensive effort in the United States and Britain was also directed toward the problem of aiming antiaircraft guns, or "fire control": the topic of the secret meeting of the National Defense Research Committee (NDRC) that George Stibitz attended, where he suggested the term *digital* for a class of devices.

The NDRC was established in June 1940. Its chair was Vannevar Bush, an MIT professor of electrical engineering who had moved to Washington, D.C., to assume the presidency of the Carnegie Institution. In the late 1930s, Bush was among those who saw a need for new types of calculating machinery to assist scientists and engineers and had the even more radical observation that such machines, when completed, would revolutionize large swaths of pure as well as applied mathematics.

While at MIT, Bush developed an analog computer called the Differential Analyzer, one of many devices that used a spinning disk to solve differential equations (power companies use a similar wheel in the meters that compute kilowatt-hours consumed by a residence). He and his students explored a variety of other mechanical and electronic devices for fire control and solving other mathematical problems, including cryptography. As early as 1938, Bush proposed a "rapid arithmetical machine" that would calculate using vacuum tubes. With his move from Cambridge to Washington coinciding with the outbreak of war in Europe in 1939, the priorities shifted. Work on the rapid arithmetical machine continued, and although its design was quite advanced, a complete working system was never completed. A bachelor's and later master's thesis by Perry Crawford, an MIT student, described what would have been a very sophisticated electronic digital computer had it been implemented.[17] Another MIT student, Claude

Shannon, had a part-time job operating the differential analyzer, and from his analysis of the relays used in it, he recognized the relationship between the simple on-off nature of relay switching and the rules of binary arithmetic. That became the basis for his master's thesis, published in 1938.[18] It has been regarded as a foundational document in the emergence of the digital age. It mirrored what Zuse had independently discovered in Berlin around the same time, and the thesis put on solid theoretical grounds what had been discovered on an ad hoc basis elsewhere. George Stibitz independently discovered this principle in 1937, and after building a breadboard circuit at his home, he went on to oversee the construction of several digital fire-control devices at Bell Labs. And long before 1937, railroads had come up with the idea of an "interlocking": a mechanical or electromechanical apparatus that ensured that in the complex switching in a rail yard, no two trains would be sent along the same track at the same time—an example of what computer scientists would later call an "exclusive-or" circuit.

As the war progressed, especially after December 1941, any project that could address the problem of the aiming and directing of guns, especially against enemy aircraft, received the highest support. Among those pressed into service was Norbert Wiener, an MIT mathematician who worked out mathematical theories about how to track a target in the presence of both electrical and other noise in the tracking system and against the ability of an enemy pilot to take evasive action. Wiener proposed building special-purpose machinery to implement his ideas, but they were not pursued. His mathematical theories turned out to have a profound impact, not only on the specific problem of fire control but on the general question of control of machinery by automatic means. In 1948 Wiener coined the term *cybernetics* and published a book under the name. It was one of the most influential books about the coming digital era, even if it did not directly address the details of digital electronic computing.[19]

Two researchers at Bell Laboratories, David Parkinson and Clarence A. Lowell, developed an antiaircraft gun director that used electronic circuits as an analog computer. The M-9 gun director was effectively used throughout the war and, in combination with the proximity fuse, neutralized the effectiveness of the German V-1 robot "buzz bomb" (because the V1 had no pilot, it could not take evasive action under fire). One interesting feature of the M-9 was that it stored equations used to compute a trajectory as wire wound around a two-dimensional camshaft, whose geometry mimicked the equation. Its most significant breakthrough was its ability to modify the aiming of the gun based on the ever-changing data fed to it by the radar tracking, an ability to direct itself toward a goal without human interaction. The success of the M-9 and other analog fire-control devices may have slowed research on the more ambitious digital designs, but the notion of self-regulation and feedback would be embodied in the software developed for digital systems as they came to fruition.

The Digital Paradigm

A central thesis of this narrative is that the digital paradigm, whose roots lay in Turing's 1936 paper, Shannon's thesis, and elsewhere, is the key to the age that defines technology today. The previous discussion of the many devices, digital or analog, special or general purpose, the different theories of computation, and the emphasis on feedback and automatic control, is not a diversion. It was out of this ferment of ideas that the paradigm emerged. While Bush's preference for analog techniques would appear to put him on the wrong side of history, that neglects the enormous influence that his students and others at MIT and Bell Laboratories had on the articulation of what it meant to be "digital." The 1940s was a decade when fundamental questions were raised about the proper role of how human beings should interact with complex control machinery. Do we construct machines that do what is technically feasible and adapt the human to their capabilities, or do we consider what humans cannot do well and try to construct machines that address those deficiencies? The answer is to do both, or a little of each, within the constraints of the existing technological base. Modern tablet computers and other digital devices do not bear much physical resemblance to the fire-control machines of the 1940s, but the questions of a human-machine interface stem from that era.

At the end of the war, Bush turned to a general look at what kinds of information processing machines might be useful in peacetime. He wrote a provocative and influential article for the *Atlantic Monthly*, "As We May Think," in which he foresaw the glut of information that would swamp science and learning if it were not controlled.[20] He suggested a machine, which he called the "Memex," to address this issue. As proposed, Memex would do mechanically what humans do poorly: store and retrieve large amounts of facts. It would also allow its human users to exploit what human do well: make connections and jump from one thread of information to another. Memex was never completed, but a vision of it, and that magazine article, had a direct link decades later to the developers of the World Wide Web. Likewise, Norbert Wiener's *Cybernetics* did not articulate the digital world as much the writings of others. The term, however, was adopted in 1982 by the science-fiction author William Gibson, who coined the term *cyberspace*—a world of bits. Wiener contributed more than just the word: his theories of information handling in the presence of a noisy environment form a large part of the foundation of the modern information-based world.

The 1940s was a decade when fundamental questions were raised about the proper role of how human beings should interact with complex control machinery. Do we construct machines that do what is technically feasible and adapt the human to their capabilities, or do we consider what humans cannot do well and try to construct machines that address those deficiencies?

The ENIAC

Much of the work described here involved the aiming of antiaircraft guns or guns mounted on ships. The aiming of large artillery also required computation, and to that end, Bush's differential analyzer was copied and heavily used to compute firing tables used in the field. Human computers, mostly women, also produced these tables. Neither method was able to keep up with wartime demand. From that need emerged a machine called the ENIAC (Electronic Numerical Integrator and Computer), unveiled to the public in 1946 at the University of Pennsylvania's Moore School of Electrical Engineering in Philadelphia. With its 18,000 vacuum tubes, the ENIAC was touted as being able to calculate the trajectory of a shell fired from a cannon faster than the shell itself traveled. That was a well-chosen example, as such calculations were the reason the army spent over a half-million dollars for the risky and unproven technique of calculating with unreliable vacuum tubes. The ENIAC used tubes for both storage and calculation, and thus could solve complex mathematical problems at electronic speeds.

The ENIAC was designed by John Mauchly and J. Presper Eckert (no relation to Wallace Eckert) at the Moore School. It represented a staggering increase in ambition and complexity over most of the ambitious computing machines already in use. It did not arise de novo. In the initial proposal to the army, Mauchly described it as an electronic version of the Bush differential analyzer, careful to stress its continuity with existing technology rather than the clean break it made. And Mauchly had visited J. V. Atanasoff in Iowa for several days in June 1941, where he most likely realized that computing with vacuum tubes at high speeds was feasible.[21] The ENIAC's design was nothing like either the Babbage or the Atanasoff machines. It used the decimal system of arithmetic, with banks of vacuum tubes that replicated the decimal wheels of an IBM tabulator. The banks of tubes were used for both calculation and storage—no separation of the two as Babbage, Zuse, and others had proposed and as is common today. The flow of numbers through the machine was patterned after the flow through the analog differential analyzer.

Of the many attributes that set the ENIAC apart, the greatest was its ability to be programmed to solve different problems. Programming was difficult and tedious. Today we click a mouse or touch an icon to call up a new program—a consequence of the stored program principle. Eckert and Mauchly designed the ENIAC to be programmable by plugging the various computing elements of it in different configurations, effectively rewiring the machine for each new problem. It was the only way to program a high-speed device until high-speed memory devices were invented. There was no point in having a device that could calculate at electronic speeds if the instructions were fed to it at mechanical speeds. Reprogramming the ENIAC to do a different job might require days, even if, once rewired, it could calculate an answer in minutes. For that reason, historians are reluctant to call the ENIAC a true "computer," a term they reserve for machines that can be flexibly reprogrammed to solve a variety of problems. But remember that the "C" in the acronym stood for "computer," a term that Eckert and Mauchly deliberately chose to evoke the rooms in which women computers operated calculating machines. The ENIAC team also gave us the term *to program* referring to a computer. Today's computers do all sorts of things besides solve mathematical equations, but it was that function for which the computer was invented and from which the machine got its name.

3

THE STORED PROGRAM PRINCIPLE

If the ENIAC, a remarkable machine, were a one-time development for the U.S. Army, it would hardly be remembered. But it is remembered for at least two reasons. First, in addressing the shortcomings of the ENIAC, its designers conceived of the stored program principle, which has been central to the design of all digital computers ever since. This principle, combined with the invention of high-speed memory devices, provided a practical alternative to the ENIAC's cumbersome programming. By storing a program and data in a common high-speed memory, not only could programs be executed at electronic speeds; the programs could also be operated on as if they were data—the ancestor of today's high-level languages compiled inside modern computers. A report that John von Neumann wrote in 1945, after he learned of the ENIAC then under construction, proved to be influential and led to several projects in the United States and Europe.[1] Some accounts called these computers "von Neumann machines," a misnomer since his report did not fully credit others who contributed to the concept. The definition of *computer* thus changed, and to an extent it remains the one used today, with the criterion of programmability now extended to encompass the internal storage of that program in high-speed memory.

From 1945 to the early 1950s, a number of firsts emerged that implemented this stored program concept. An experimental stored program computer was completed in Manchester, United Kingdom, in 1948 and carried out a rudimentary demonstration of the concept. It has been claimed to be the first stored program electronic computer in the world to execute a program. The following year, under the direction of Maurice Wilkes in Cambridge, a computer called the EDSAC (Electronic Delay Storage Automatic Computer) was completed and began operation. Unlike the Manchester "baby," it was a fully functional and useful stored program computer. Similar American efforts came to fruition shortly after, among them the SEAC, built at the U.S. National Bureau of Standards and operational in 1950. The IBM Corporation also built an electronic computer, the SSEC, which contained a memory device that stored instructions that could be modified during a calculation (hence the name, which stood for "selective sequence electronic calculator"). It was not designed along the lines of the ideas von Neumann had expressed, however. IBM's Model 701, marketed a few years later, was that company's first such product.

The second reason for placing the ENIAC at such a high place in history is that its creators, J. Presper Eckert and John Mauchly, almost alone in the early years sought to build and sell a more elegant follow-on to the ENIAC for commercial applications. The ENIAC was conceived, built, and used for scientific and military applications. The UNIVAC, the commercial computer, was conceived and marked as a general-purpose computer suitable for any application that one could program it for. Hence the name: an acronym for "universal automatic computer."

Eckert and Mauchly's UNIVAC was well received by its customers. Presper Eckert, the chief engineer, designed it conservatively, for example, by carefully limiting the loads on the vacuum tubes to prolong their life. That made the UNIVAC surprisingly reliable in spite of its use of about 5,000 vacuum tubes. Sales were modest, but the technological breakthrough of a commercial electronic computer was large. Eckert and Mauchly founded a company, another harbinger of the volatile computer industry that followed, but it was unable to survive on its own and was absorbed by Remington Rand in 1950. Although not an immediate commercial success, the UNIVAC made it clear that electronic computers were eventually going to replace the business and scientific machines of an earlier era. That led in part to the decision by the IBM Corporation to enter the field with a large-scale electronic computer of its own, the IBM 701. For its design, the company drew on the advice of von Neumann, whom it hired as a consultant. Von Neumann had proposed using a special-purpose vacuum tube developed at RCA for memory, but they proved difficult to produce in quantity. The 701's memory units were special-purpose modified cathode ray tubes, which had been invented in England and used on the Manchester computer.

IBM initially called the 701 a "defense calculator," an acknowledgment that its customers were primarily aerospace and defense companies with access to large amounts of government funds. Government-sponsored military agencies were also among the first customers for the UNIVAC, now being marketed by Remington Rand. The 701 was optimized for scientific, not business, applications, but it was a general-purpose computer. IBM also introduced machines that were more suitable for business applications, although it would take several years before business customers could obtain a computer that had the utility and reliability of the punched card equipment they had been depending on.

By the mid-1950s, other vendors joined IBM and Remington Rand as U.S. and British universities were carrying out advanced research on memory devices, circuits, and, above all, programming. Much of this research was supported by the U.S. Air Force, which contracted with IBM to construct a suite of large vacuum tube computers to be used for air defense. The system, called "SAGE" (Semi-Automatic Ground Environment), was controversial: it was built to compute the paths of possible incoming Soviet bombers, and to calculate a course for U.S. jets to intercept them. But by the time the SAGE system was running, the ballistic missile was well under development, and SAGE could not aid in ballistic missile defense. Nevertheless, the money spent on designing and building multiple copies of computers for SAGE was beneficial to IBM, especially in developing mass memory devices that IBM could use for its commercial computers.[2] There was a reason behind the acronym for the system: *semiautomatic* was deliberately chosen to assure everyone that in this system, a human being was always in the loop when decisions were made concerning nuclear weapons. No computer was going to start World War III by itself, although there were some very close calls involving computers, radars, and other electronic equipment during those years of the Cold War. In some cases the false alarms were caused by faulty computer hardware, in another by poorly written software, and also when a person inadvertently loaded a test tape that simulated an attack, which personnel interpreted to mean a real attack was underway.[3] Once again, the human-machine interface issue appeared.

The Mainframe Era, 1950–1970

The 1950s and 1960s were the decades of the mainframe, so-called because of the large metal frames on which the computer circuits were mounted. The UNIVAC inaugurated the era, but IBM quickly came to dominate the industry in the United States, and it made significant inroads into the European market as well.

In the 1950s, mainframes used thousands of vacuum tubes; by 1960 these were replaced by transistors, which consumed less power and were more compact. The machines were still large, requiring their own air-conditioned rooms and a raised floor, with the connecting cables routed beneath it. Science fiction and popular culture showcased the blinking lights of the control panels, but the banks of spinning reels of magnetic tape that stored the data formerly punched onto cards were the true characteristic of the mainframe. The popular image of these devices was that of "giant brains," mainly because of the tapes' uncanny ability to run in one direction, stop, read or write some data, then back up, stop again, and so on, all without human intervention. Cards did not disappear, however: programs were punched onto card decks, along with accompanying data, both of which were transferred to reels of tape. The decks were compiled into batches that were fed into the machine, so that the expensive investment was always kept busy. Assuming a program ran successfully, the results would be printed on large, fan-folded paper, with sprocket holes on both sides. In batch operation, the results of each problem would be printed one after another; a person would take the output and separate it, then place the printed material in a mail slot outside the air-conditioned computer room for those who needed it. If a program did not run successfully, which could stem from any number of reasons, including a mistake in the programming, the printer would spew out error messages, typically in a cryptic form that the programmer had to decode. Occasionally the computer would dump the contents of the computer's memory onto the printed output, to assist the programmer. *Dump* was very descriptive of the operation. Neither the person whose problem was being solved nor the programmer (who typically were different people) had access to the computer.

Because these computers were so expensive, a typical user was not allowed to interact directly with them. Only the computer operator had access: he (and this job was typically staffed by men) would mount or demount tapes, feed decks of cards into a reader, pull results off a printer, and otherwise attend to the operation of the equipment. A distaste for batch operations drove enthusiasts in later decades to break out of this restricted access, at first by using shared terminals connected directly to the mainframe, later by developing personal computers that one could use (or abuse) as one saw fit. Needless to say, word processing,

Every machine requires a set of procedures to get it to do what it was built to do. Only computers elevate those procedures to a status equal to that of the hardware and give them a separate name.

code breaking). A person skilled in mathematical analysis would lay out the steps needed to solve a problem, translate those steps into a code that was intrinsic to the machine's design, punch those codes into paper tape or its equivalent, and run the machine. Programming was straightforward, although not as easy as initially envisioned: programmers soon found that their codes contained numerous "bugs," which had to be rooted out.[4]

Once general-purpose computers like the UNIVAC (note the name) began operation, the importance of programming became clearer. At this moment, a latent quality of the computer emerged to transform the activity of programming. When von Neumann and the ENIAC team conceived of the stored program principle, they were thinking of the need to access instructions at high speed and to have a simple design of the physical memory apparatus to accommodate a variety of problems. They also noted that the stored program could be modified, in the restricted sense that an arithmetic operation could address a sequence of memory locations, with the address of the memory incremented or modified according to the results of a calculation. Through the early 1950s, that was extended: a special kind of program, later called a *compiler*, would take as its input instructions written in a form that was familiar to human programmers and easy to grasp , and its output would be another program, this one written in the arcane codes that the hardware was able to decode. That concept is at the root of modern user interfaces, whether they use a mouse to select icons or items on a screen, a touch pad, voice input—anything.

The initial concept was to punch common sequences, such as computing the sine of a number, onto a tape or deck of cards, and assemble, or compile, a program by taking these precoded sequences, with custom-written code to knit them to one another. If those sequences were stored on reels of tape, the programmer had only to write out a set of commands that called them up at the appropriate time, and these commands could be written in a simpler code that did not have to keep track of the inner minutiae of the machine. The concept was pioneered by the UNIVAC team, led by Grace Murray Hopper (1906–1992), who had worked with Howard Aiken at Harvard during World War II (see figure 3.1). The more general concept was perhaps first discovered by two researchers, J. H. Laning and N. Zierler for the Whirlwind, an air force–sponsored computer and the ancestor of the SAGE air defense system, at the MIT in the early 1950s. Unlike the UNIVAC compilers, this system worked much as modern compilers do: it took as its input commands that a user entered, and it generated as output fresh and novel machine code that executed those commands and kept track of storage locations, handled repetitive loops, and did other housekeeping chores. Laning and Zierler's "algebraic system" was the first step toward such a compiler: it took commands a user typed in

Figure 3.1 Grace Murray Hopper and students at the control panel of a UNIVAC computer in the mid-1950s. Hopper helped design systems that made it easier to program the UNIVAC, and she devoted her long career in computing to making them more accessible and useful to nonspecialists. (Credit: Grace Murray Hopper/Smithsonian Institution).

familiar algebraic form and translated them into machine codes that Whirlwind could execute.

The Whirlwind system was not widely adopted, however, as it was perceived to be wasteful of expensive computer resources. Programming computers remained in the hands of a priesthood of specialists who were comfortable with arcane machine codes. That had to change as computers grew more capable, the problems they were asked to solve grew more complex, and researchers developed compilers that were more efficient. The breakthrough came in 1957 with Fortran, a programming language IBM introduced for its Model 704 computer, and a major success. (It was also around this time that these codes came to be called "languages," because they shared many, though not all, characteristics with spoken languages.) Fortran's syntax—the choice of symbols and the rules for using them—was close to ordinary algebra, which made it familiar to engineers. And there was less of a penalty in overhead: the Fortran compiler generated machine code that was as efficient and fast as code that human beings wrote. IBM's dominant market position also played a role in its success. People readily embraced a system that hid the details of a machine's inner workings, leaving them free to concentrate on solving their own, not the machine's, problems.

Fortran's success was matched in the commercial world by COBOL (Common Business Oriented Language). COBOL owed its success to the U.S. Department of Defense, which in May 1959 convened a committee to address the question of developing a common business language; that meeting was followed by a brief and concentrated effort to produce specifications for the language. As soon as those were published, manufacturers set out writing compilers for their respective computers. The next year the U.S. government announced that it would not purchase or lease computer equipment unless it could handle COBOL. As a result, COBOL became one of the first languages to be standardized to a point where the same program could run on different computers, from different vendors, and produce the same results. The first recorded instance of that milestone occurred in December 1960, when the same program (with a few minor changes) ran on a UNIVAC II and an RCA 501.

As with Fortran, the backing of a powerful organization—in this case, the Department of Defense—obviously helped win COBOL's acceptance. Whether the language itself was well designed and capable is still a matter of debate. It was designed to be readable, with English commands like "greater than" or "equals" replacing the algebraic symbols >, =, and the others. The commands could even be replaced by their equivalent commands in a language other than English. Proponents argued that this design made the program easier to read and understand by managers who used the program but had little to do with writing it. That belief was misplaced: the programs were still hard to follow, especially years later when someone else tried to modify them. And programmers tended to become comfortable with shorter codes if they used them every day. Programming evolved in both directions: for the user, English commands such as "print" or "save" persisted but were eventually replaced by symbolic actions on icons and other graphic symbols. Programmers gravitated toward languages such as C that used cryptic codes but also allowed them to get the computer to perform at a high level of efficiency.

As commercial computers took on work of greater complexity, another type of program emerged that replaced the human operators who managed the flow of work in a punched card installation: the operating system. One of the first was used at the General Motors Research Laboratories around 1956, and IBM soon followed with a system it called JCL (Job Control Language). JCL consisted of cards punched with codes to indicate to the computer that the deck of cards that followed was written in Fortran or another language, or that the deck contained data, or that another user's job was coming up. Systems software carried its own overhead, and if it was not well designed, it could cripple the computer's ability to work efficiently. IBM struggled with developing operating systems, especially for its ambitious System/360 computer series, introduced in 1964. IBM's struggles were replaced in the 1980s and 1990s by Microsoft's, which produced windows-based systems software for personal computers. As a general rule, operating systems tend to get more and more complex with each new iteration, and with that comes increasing demands on the hardware. From time to time, the old inefficient systems are pushed aside by new systems that start out lean, as with the current Linux operating system, and the lean operating systems developed for smart phones (although these lean systems, like their predecessors, get more bloated with each release) (see figure 3.2).

The Transistor

A major question addressed in of this narrative is to what extent the steady advance of solid-state electronics has driven history. Stated in the extreme, as modern technology crams ever more circuits onto slivers of silicon, new devices and applications appear as ripe fruit falling from a tree. Is that too extreme? We saw how vacuum tubes in place of slower mechanical devices were crucial to the invention of the stored program digital computer. The phenomenon was repeated again, around 1960, by the introduction of the solid-state transistor as a replacement for tubes.

The vacuum tubes used in the ENIAC offered dramatically higher speeds: the switching was done by electrons,

Figure 3.2 A large IBM System/360 installation. Note the control panel with its numerous lights and buttons (center) and the rows of magnetic tape drives, which the computer used for its bulk memory. Behind the control panel in the center is a punched card reader with a feed of fan-folded paper; to the right is a printer and a video display terminal. (Credit: IBM Corporation)

excited by a hot filament, moving at high speeds within a vacuum. Not all computer engineers employed them; Howard Aiken and George Stibitz, for example, were never satisfied with tubes, and they preferred devices that switched mechanically. Like the familiar light bulb from which they were descended, tubes tended to burn out, and the failure of one tube among thousands would typically render the whole machine inoperative until it was replaced (so common was this need for replacement that tubes were mounted in sockets, not hardwired into the machine as were other components). Through the early twentieth century, physicists and electrical engineers sought ways of replacing the tube with a device that could switch or amplify currents in a piece of solid material, with no vacuum or hot filament required. What would eventually be called solid-state physics remained an elusive goal, however, as numerous attempts to replace active elements of a tube with compounds of copper or lead, or the element germanium, foundered.[5] It was obvious that a device to replace vacuum tubes would find a ready market. The problem was that the theory of how these devices operated lagged. Until there was a better understanding of the quantum physics behind the phenomenon, making a solid-state replacement for the tube was met with frustration. During World War II, research in radio detection of enemy aircraft (radar) moved into the realm of very high and ultrahigh frequencies (VHF, UHF), where traditional vacuum tubes did not work well. Solid-state devices were invented that could rectify radar frequencies (i.e., pass current in only one direction). These were built in quantity, but they could not amplify or switch currents.

After the war, a concentrated effort at Bell Laboratories led to the invention of the first successful solid-state switching or amplifying device, which Bell Labs christened the "transistor." It was invented just before Christmas 1947. Bell Labs publicly demonstrated the invention, and the *New York Times* announced it in a modest note in the "The News of Radio" column in on July 1, 1948 —in one of the greatest understatements in the history of technology. The public announcement was crucial because Bell Labs was part of the regulated telephone monopoly; in other words, the revolutionary invention was not classified or retained by the U.S. military for weapons use. Instead it was to be offered to commercial manufacturers. The three inventors, William Shockley, Walter Brattain, and John Bardeen, shared the 1956 Nobel Prize in physics for the invention.

It took over a decade before the transistor made a significant effect on computing. Much of computing history has been dominated by the impact of Moore's law, which describes the doubling of transistors on chips of silicon since 1960. However, we must remember that for over a decade prior to 1960, Moore's law had a slope of zero: the number of transistors on a chip remained at one. Bringing the device to a state where one could place more than one of them on a piece of material, and get them to operate reliably and consistently, took a concentrated effort. The operation of a transistor depends the quantum states of electrons as they orbit a positively charged nucleus of an atom. By convention, current is said to flow from a positive to a negative direction, although in a conductor such as a copper wire, it is the negatively charged electrons that flow the other way. What Bell Labs scientists discovered was that in certain materials, called semiconductors, current could also flow by moving the *absence* of an electron in a place where it normally would be: positive charges that the scientists called holes. Postwar research concentrated specifically on two elements that were amenable to such manipulation: germanium and silicon. The 1947 invention used germanium, and it worked by having two contacts spaced very near one another, touching a substrate of material. It worked because its inventors understood the theory of what was happening in the device. Externally it resembled the primitive "cat's whisker" detector that amateur radio enthusiasts used through the 1950s—a piece of fine wire carefully placed by hand on a lump of galena—hardly a device that could be the foundation of a computer revolution.[6]

From 1947 to 1960 the foundations of the revolution in solid state were laid. First, the point-contact device of 1947 gave way to devices in which different types of semiconductor material were bonded rigidly to one another. Methods were developed to purify silicon and germanium by melting the material in a controlled fashion and growing a crystal whose properties were known and reproducible. Precise amounts of impurities were deliberately introduced, to give the material the desired property of moving holes (positive charges, or P-type) or electrons (negative charges, or N-type), a process called doping. The industry managed a transition from germanium to silicon, which was harder to work but has several inherent advantages. Methods were developed to lay down the different material on a substrate by photolithography. This last step was crucial: it allowed miniaturizing the circuits much like one could microfilm a book. The sections of the transistor could be insulated from one another by depositing layers of silicon oxide, an excellent insulator.

Manufacturers began selling transistorized computers beginning in the mid-1950s. Two that IBM introduced around 1959 marked the transition to solid state. The Model 1401 was a small computer that easily fit into the existing punched card installations then common among commercial customers. Because it eased the transition to electronic computers for many customers, it became one of IBM's best-selling products, helping to solidify the company's dominant position in the industry into the 1980s. The IBM 7090 was a large mainframe that had the same impact among scientific, military, and aerospace customers.

It had an interesting history: IBM bid on a U.S. Air Force contract for a computer to manage its early-warning system then being built in the Arctic to defend the United States from a Soviet missile attack (a follow-on to SAGE). IBM had just introduced Model 709, a large and capable vacuum tube mainframe, but the air force insisted on a transistorized computer. IBM undertook a heroic effort to redesign the Model 709 to make it the transistorized Model 7090, and it won the contract.

The Minicomputer

The IBM 7090 was a transistorized computer like the radio was a "wireless" telegraph or the automobile was a "horseless" carriage: it retained the mental model of existing computers, only with a new and better base technology. What if one started with a clean sheet of paper and designed a machine that took advantage of the transistor for what it was, not for what it was replacing?

In 1957, Ken Olsen, a recent graduate of MIT, and Harlan Anderson, who had been employed at MIT's Digital Computer Laboratory, did just that.[7] Both had been working at the Lincoln Laboratory, a government-sponsored lab outside Boston that was concerned with problems of air defense. While at Lincoln they worked on the TX-0, a small computer that used a new type of transistor manufactured by the Philco Corporation. The TX-0 was built to test the feasibility of that transistor for large-scale computers. It not only demonstrated that; it also showed that a small computer, using high-speed transistors, could outperform a large mainframe in certain applications. Olsen and Anderson did not see the path that computers would ultimately take, and Olsen in particular would be criticized for his supposed lack of foresight. But they opened a huge path to the future: from the batch-oriented mainframes of IBM to the world we know today: a world of small, inexpensive devices that could be used interactively.

With $70,000 of venture capital money—another innovation—Olsen and Anderson founded the Digital Equipment Corporation (DEC, pronounced as an acronym), and established a plant at a nineteenth-century woolen mill in suburban Maynard, Massachusetts. Their initial product was a set of logic modules based on high-performance transistors and associated components. Their ultimate goal was to produce a computer, but they avoided that term because it implied going head-to-head with IBM, which most assumed was a foolish action. They chose instead the term *programmed data processor.* In 1960, DEC announced the PDP-1, which in the words of a DEC historian, "brought computing out of the computer room and into the hands of the user."[8] The company sold the prototype PDP-1 to the Cambridge research firm Bolt Beranek and Newman (BBN), where it caught the eye of one of its executives, psychologist J. C. R. Licklider. We return to Licklider's role in computing later, but in any event, the sale was fortuitous (see figure 3.3).

In April 1965 DEC introduced another breakthrough computer, the PDP-8. Two characteristics immediately set it apart. The first was its selling price: $18,000 for a basic system. The second was its size: small enough to be mounted alongside standard equipment racks in laboratories, with no special power or climate controls needed. The machine exhibited innovations at every level. Its logic circuits took advantage of the latest in transistor technology. The circuits themselves were incorporated into standard modules, which in turn were wired to one another by an automatic machine that employed wire wrap, a technique developed at Bell Labs to execute complex wiring that did not require hand soldering. The modest size and power requirements have already been mentioned. The overall architecture, or functional design, was also innovative. The PDP-8 handled only 12 bits at a time, far fewer than mainframes. That lowered costs and increased speed as long as it was not required to do extensive numerical calculations. This architecture was a major innovation from DEC and signified that the PDP-8 was far more than a vacuum tube computer designed with transistors.

At the highest level, DEC had initiated a social innovation as significant as the technology. DEC was a small company and did not have the means to develop specific

Figure 3.3 Engineers at the Digital Equipment Corporation introducing their large computer, the PDP-6, in 1964. Although DEC was known for its small computers, the PDP-6 and its immediate successor, the PDP-10, were large machines that were optimized for time-sharing. The PDP-6 had an influence on the transition in computing from batch to interactive operation, to the development of artificial intelligence, and to the ARPANET. Bill Gates and Paul Allen used a PDP-10 at Harvard University to write some of their first software for personal computers. *Source*: Digital Equipment Corporation—now Hewlett-Packard. Credit: Copyright© Hewlett-Packard Development Company, L.P. Reproduced with permission.

applications for its customers, as IBM's sales force was famous for doing. Instead it encouraged the customers themselves to develop the specialized systems hardware and software. It shared the details of the PDP-8's design and operating characteristics and worked with customers to embed the machines into controllers for factory automation, telephone switching, biomedical instrumentation, and other uses: a restatement in practical terms of the general-purpose nature of the computer as Turing had outlined in 1936. The computer came to be called a "mini," inspired by the Morris Mini automobile, then being sold in the United Kingdom, and the short skirts worn by young women in the late 1960s. The era of personal, interactive computing had not yet arrived, but it was coming.

The Convergence of Computing and Communications

While PDP-8s from Massachusetts were blanketing the country, other developments were occurring that would ultimately have an equal impact, leading to the final convergence of computing with communications. In late November 1962, a chartered train was traveling from the Allegheny Mountains of Virginia to Washington, D.C. The passengers were returning from a conference on "information system sciences," sponsored by the U.S. Air Force and the MITRE Corporation, a government-sponsored think tank. The conference had been held at the Homestead Resort in the scenic Warm Springs Valley of Virginia. The attendees had little time to enjoy the scenery or take in the healing waters of the springs, however. A month earlier, the United States and Soviet Union had gone to the brink of nuclear war over the Soviet's placement of missiles in Cuba. Poor communications, not only between the two superpowers but between the White House, the Pentagon, and commanders of ships at sea, were factors in escalating the crisis.

Among the passengers on that train was J. C. R. Licklider, who had just moved from Bolt Beranek and Newman to become the director of the Information Processing Techniques Office (IPTO) of a military agency called the Advanced Research Projects Agency (ARPA). That agency (along with NASA, the National Aeronautics and Space Administration) was founded in 1958, in the wake of the Soviet's orbiting of the *Sputnik* earth satellites. Licklider now found himself on a long train ride in the company of some of the brightest and most accomplished computer scientists and electrical engineers in the country. The conference in fact had been a disappointment. He had heard a number of papers on ways digital computers could improve military operations, but none of the presenters, in his view, recognized the true potential of computers in military—or civilian—activities. The train ride gave him a chance to reflect on this potential and to share it with his colleagues.

Licklider had two things that made his presence among that group critical. The first was money: as director of IPTO, he had access to large sums of Defense Department funds and a free rein to spend them on projects as he saw fit. The second was a vision: unlike many of his peers, he saw the electronic digital computer as a revolutionary device not so much because of its mathematical abilities but because it could be used to work in symbiosis—his favorite term—with human beings. Upon arrival in Washington, the passengers dispersed to their respective homes. Two days later, on the Friday after Thanksgiving, Robert Fano, a professor at MIT, initiated talks with his supervisors to start a project based on the discussions he had had on the train two days before. By the following Friday, the outline was in place for Project MAC, a research initiative at MIT to explore "machine-aided cognition," by configuring computers to allow "multiple access"—a dual meaning of the acronym. Licklider, from his position at ARPA, would arrange for the U.S. Office of Naval Research to fund the MIT proposal with an initial contract of around $2.2 million. Such an informal arrangement might have raised eyebrows, but that was how ARPA worked in those days.

Licklider's initial idea was to start with a large mainframe and share it among many users simultaneously. The concept, known as time-sharing, was in contrast to the sequential access of batch operation. If it was executed properly, users would be unaware that others were also using the machine because the computer's speeds were far higher than the response of the human brain or the dexterity of human fingers. The closest analogy is a grandmaster chess player playing a dozen simultaneous games with less capable players. Each user would have the illusion (a word used deliberately) that he or she had a powerful computer at his personal beck and call.

Licklider and most of his colleagues felt that time-sharing was the only practical way to evolve computers to a point where they could serve as direct aids to human cognition. The only substantive disagreement came from Wes Clark, who had worked with Ken Olsen at Lincoln Laboratories on transistorized computers. Clark felt this approach was wrong: he argued that it would be better to evolve small computers into devices that individuals could use directly. He was rebuffed by electrical and computer engineers but found a more willing audience among medical researchers, who saw the value of a small computer that would not be out of place among the specialized equipment in their laboratories. With support from the National Institutes of Health, he assembled a team that built a computer called the "LINC" (from Lincoln Labs), which he demonstrated in 1962. A few were built, and later they were combined with the Digital Equipment Corporation

PDP-8. It was a truly interactive, personal computer for those few who had the privilege of owning one. The prevailing notion at the time, however, was still toward time-sharing large machines; personal computing would not arrive for another decade.[9] The dramatic lowering of the cost of computing that came with advances in solid-state electronics made that practical, and that was exploited by young people like Steve Jobs and Bill Gates who were not part of the academic research environment.

Clark was correct in recognizing the difficulties of implementing time-sharing. Sharing a single large computer among many users with low-powered terminals presented many challenges. While at BBN, Licklider's colleagues had configured the newly acquired PDP-1 to serve several users at once. Moving that concept to large computers serving large numbers of users, beginning with an IBM and moving to a General Electric mainframe at MIT, was not easy. Eventually commercial time-sharing systems were marketed, but not for many years and only after the expenditure of a lot of development money. No matter: the forces that Licklider set in motion on that train ride transformed computing.

Time-sharing was the spark that set other efforts in motion to join humans and computers. Again with Licklider's support, ARPA began funding research to interact with a computer using graphics and to network geographically dispersed mainframes to one another.[10] Both were crucial to interactive computing, although it was ARPA's influence on networking that made the agency famous. A crucial breakthrough came when ARPA managers learned of the concept of *packet switching*, conceived independently in the United Kingdom and the United States. The technique divided a data transfer into small chunks, called packets, which were separately addressed and sent to their destination and could travel over separate channels if necessary. That concept was contrary to all that AT&T had developed over the decades, but it offered many advantages over classical methods of communication, and remains the technical backbone of the Internet to this day.

The first computers of the "ARPANET" were linked to one another in 1969; by 1971 there were fifteen computers, many of them from the Digital Equipment Corporation, connected to one another. The following year, ARPA staged a demonstration at a Washington, D.C., conference. The ARPANET was not the sole ancestor of the Internet as we know it today: it was a military-sponsored network that lacked the social, political, and economic components that together comprise the modern networked world. It did not even have e-mail, although that was added soon. The demonstration did show the feasibility of packet switching, overcoming a lot of skepticism from established telecommunications engineers who believed that such a scheme was impractical. The rules for addressing and routing packets, which ARPA called protocols, worked.

ARPA revised them in 1983 to a form that allowed the network to grow in scale. These protocols, called Transmission Control Protocol/Internet Protocol (TCP/IP) remain in use and are the technical foundation of the modern networked world.

4

THE CHIP AND SILICON VALLEY

After a decade of slow but steady progress in transistor development, a breakthrough occurred in the early 1960s: simultaneously inventors in Texas and California devised a way of placing multiple transistors and other devices on a single chip of silicon. That led rapidly to circuits that could store ever-increasing amounts of data—the storage component of computing. The integrated circuit (IC) was a breakthrough, but it was also part of a long evolutionary process of miniaturization of electronic circuits. Well before the invention of the transistor, manufacturers of hearing aids sought ways to make their products small and light enough to be worn on the body and even concealed if possible.

That impetus to miniaturize vacuum tubes had a direct influence on one of the most famous secret weapons of World War II, the Proximity Fuze—a device that used radio waves to detonate a shell at a distance calculated to be most effective in destroying an enemy aircraft, but not requiring a direct hit. Combined with the analog gun directors mentioned earlier, the Fuze gave the Allies a potent weapon, especially against the German V-1 buzz bomb. The Proximity Fuze had to fit inside a shell ("about the size of an ice-cream cone") and be rugged enough to withstand the shock and vibration of firing.[1] From that work came not only rugged miniature tubes, but also the printed circuit: a way of wiring the components of the device by laying down conducting material on a flat slab of insulating material rather than connecting the components by actual wires. Descendants of these printed circuits can be found inside almost all modern digital devices. The concept itself, of printing a circuit, would have much in common with the invention of the integrated circuit itself decades later.

One characteristic of computing that is most baffling to the layperson is how such astonishing capabilities can arise from a combination of only a few basic logical circuits: the AND, OR, NOT circuits of logic, or their mathematical equivalents: the addition, multiplication, or negation of the digits 1 and 0. The answer is that these circuits must be aggregated in sufficient numbers: a few dozen to perform simple arithmetic, a few hundred to perform more complex calculations, tens of thousands to create a digital computer, millions or billions to store and manipulate images, and so on. Besides the active components like vacuum tubes or transistors, a computer also requires many passive devices such as resistors, diodes, and capacitors. The key element of computer design, software as well as hardware, is to manage the complexity from the lower levels of logical circuits to ever-higher levels that nest above one another. One may compare this to the number of neurons in the brains of animals, from the flatworm, to a cat, to *Homo sapiens*, although the history of artificial intelligence research has shown that comparing a human brain to a computer can distort as much as clarify. A better analogy might be to compare the complexity of a computer chip to the streets of a large metropolis, like Chicago, which can support amenities not found in smaller cities: for example, symphony orchestras, major league sports teams, and international airports.

In designing the ENIAC, Eckert and Mauchly recognized the need to manage complexity, for which they designed standard modules containing a few dozen tubes and other components. These performed a basic operation of adding and storing decimal numbers. If a module failed, it could be quickly replaced by a spare. In the early 1960s, IBM developed what it called the standard modular system of basic circuits, mounted on a printed circuit board about the size of a playing card, for its mainframes. The first products from the Digital Equipment Corporation were also logic modules, which performed complex operations likely to be used in a digital computing system. Those who designed computers without such modularity found

The key element of computer design—software as well as hardware—is to manage the complexity from the lower levels of logical circuits to ever-higher levels that nest above one another. One may compare this to the number of neurons in the brains of animals, from the flatworm, to a cat, to *Homo sapiens*, although the history of artificial intelligence research has shown that comparing a human brain to a computer can distort as much as clarify.

their systems almost impossible to maintain or troubleshoot—the systems were literally "haywire."

The Invention of the Integrated Circuit

The next logical step in this process was to place all of the elements of one of those modules on a single chip of material, either germanium or silicon. That could not happen until the transistor itself progressed from the crude device of 1947 to a reliable device that emerged around 1959. Another step was required, and that was to understand how to integrate the transistors with the passive components, such as resistors and capacitors, on the same piece of material. Passive components were cheap (they cost only a few pennies each) and rugged. Why make them out of the same expensive material as the transistors?

The reason is that making the passive devices out of germanium or silicon meant that an entire logic circuit could be fashioned on a single chip, with the interconnections built in. That required a way of depositing circuit paths on the chip and insulating the various elements from one another. The conceptual breakthrough was really the first step: to consider the circuit as a whole, not as something made of discrete components. Jack Kilby, working at Texas Instruments in Dallas, took that step in 1958. Before joining Texas Instruments, Kilby had worked at a company named Centrallab in Milwaukee, an industry leader in printed circuits. He moved to Texas Instruments, which at the time was working on a government-funded project, Micro-Module, that involved depositing components on a ceramic wafer. Kilby did not find this approach cost-effective, although IBM would later use something like it for its mainframes. In summer 1958, Kilby came to the idea of making all the components of a circuit out of the same material. He first demonstrated the feasibility of the idea by building an ordinary circuit of discrete components, but all of them, including its resistors and capacitors, were made of silicon instead of the usual materials. In September he built another circuit, an oscillator, and this time all components were made from a single thin wafer of germanium. Fine gold wires connected the elements on the wafer to one another. In early 1959 he applied for a patent, which was granted in 1964.

Robert Noyce was working at Fairchild Semiconductor in Mountain View, California, when he heard of Kilby's invention. In January 1959 he described in his lab notebook a scheme for doing the same thing, but with a piece of silicon. One of his coworkers, Jean Hoerni, had literally paved the way by developing a process for making silicon transistors that was well suited for mass production. He called it the planar process. As the name implies, it produced transistors that were flat (other techniques required raised metal lines or, in Kilby's invention, wires attached to the surface). The process was best suited to silicon, where layers of silicon oxide could be used to isolate one device from another. Noyce applied for a patent in July 1959, a few months after Kilby (see figure 4.1). Years later the courts adjudicated the dispute over the competing claims, giving Kilby and Noyce, and their respective companies, a share of the credit and claims. For his work, Kilby was awarded the 2000 Nobel Prize in Physics. Noyce had passed away in 1990 at the age of sixty-two; had he lived, he no doubt would have shared the prize.[2]

The invention of the IC had two immediate effects. The first came from the U.S. aerospace community, which always had a need for systems that were small and lightweight and, above all, reliable. Developers of guided and ballistic missiles were repeatedly embarrassed by launch failures, later traced to the failure of a simple electronic component costing at most a few dollars. And as systems became more complex, the probability of wiring errors introduced by the human hand increased, no matter how carefully the assembly process was organized. In the late 1950s the U.S. Air Force was involved with the design of the guidance system for the Minuteman solid-fueled ballistic missile, where reliability, size, and weight were critical. From the Minuteman and related projects came the critical innovation of the "clean room," where workers wore gowns to keep dust away from the materials they were working with, and air was filtered to a degree not found in the cleanest hospital. At every step of the production of every electronic component

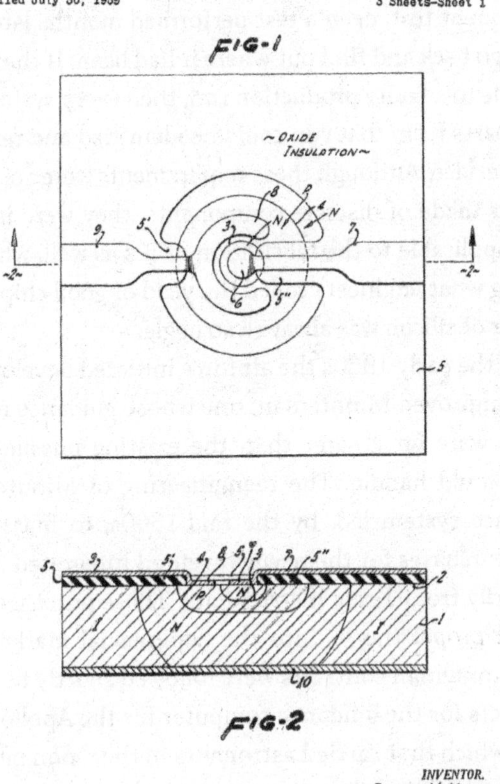

Figure 4.1 Patent for the integrated circuit, as invented by Robert Noyce of Fairchild Semiconductor.

used in Minuteman, a log was kept that spelled out exactly what was done to the part, and by whom. If a part failed a subsequent test, even a test performed months later, one could go back and find out where it had been. If the failure was due to a faulty production run, then every system that used parts from that run could be identified and removed from service. Although these requirements were applied to circuits made of discrete components, they were immediately applicable to the fabrication of ICs as well, where obtaining what engineers called the yield of good chips from a wafer of silicon was always a struggle.

In the early 1960s the air force initiated development of an improved Minuteman, one whose guidance requirements were far greater than the existing missile's computer could handle. The reengineering of Minuteman's guidance system led, by the mid-1960s, to massive air force purchases for the newly invented Integrated Circuit, primarily from Texas Instruments. Those purchases that helped propel the IC into the commercial marketplace. The Minuteman contracts were followed shortly by NASA contracts for the guidance computer for the Apollo spacecraft, which first carried astronauts to the moon between 1969 and 1972. It is no exaggeration to say that in the mid-1960s, the majority of all the ICs in the world could be found in either the guidance systems of Minuteman intercontinental ballistic missiles or in the command and lunar modules of the Apollo spacecraft.

Route 128 and Silicon Valley

Invention of the IC led to a rapid acceleration of minicomputers. The first model of the PDP-8 used discrete components, but later models used ICs, and the Digital Equipment Corporation followed with other products, notably the PDP-11, with capabilities reaching into the mainframe's market. DEC was soon joined by a host of competing minicomputer companies, which found it relatively easy to enter the market with designs based on the availability of standardized logic chips offered by Fairchild, Texas Instruments, and their brethren. Their entry was helped by a particular product offered by Fairchild: a circuit that could be used as a computer's memory in place of the magnetic cores then in common use. Cores—small doughnut-shaped pieces of magnetic material with wires threaded through them—were compact but typically had to be hand assembled, whereas IC memories could be produced along with the logic circuits. And their power needs, size, and electrical characteristics fit well on the printed circuit boards that made up a computer.

Many of these minicomputer companies were located along Route 128 in the Boston suburbs, but others were located near Fairchild in the Santa Clara Valley below San Francisco. Don Hoefler, a local journalist, renamed the region "Silicon Valley" in 1971, and it has been known by that name ever since. Fairchild was the leading company, but from the beginning of its involvement with the IC, many of its best employees began leaving the company to found rivals nearby. A running joke in Silicon Valley was that an engineer might find himself with a new job if he accidently turned into the wrong parking lot in the morning. The exodus of Fairchild employees was poetic justice: Fairchild itself was founded by defectors from a company founded in Palo Alto by William Shockley, one of the inventors of the transistor. Although disruptive in the short term, this fluidity ultimately contributed to the success of Silicon Valley as a dynamo of innovation.

Among the Fairchild spin-offs was a company called Intel, founded in 1968 by Robert Noyce, the IC coinventor, and Gordon Moore, who in 1965 observed the rapid doubling of the capacity of semiconductor memory chips. They were soon joined by Andrew Grove, also a Fairchild alumnus. Intel scored a major success in 1970 with its introduction of a memory chip, the 1103, that stored about 1,000 bits of information (128 "bytes," where a byte is defined as 8 bits). With that announcement, core memories soon became obsolete. Intel's emphasis on a memory chip was no accident and pointed to a potential problem with the invention: one could construct a complex circuit on a single chip of silicon, but the more complex the circuit, the more specialized its function, and hence the narrower its market. That has been a fundamental dilemma of mass production ever since Henry Ford tried to standardize his

Model T. Memory chips avoided this problem, since the circuits were regular and filled a need for mass storage that every computer had. But for other circuits the problem remained. We look at how it was solved in the next chapter.

IBM's System/360

In the midst of this revolution in semiconductor electronics, IBM introduced a new line of mainframe computers that transformed the high end as much as these events transformed the low end of computing. In 1964 IBM announced the System/360 line of mainframes. The name implied that the machines would address the full circle of scientific and business customers, who previously would have bought or leased separate lines of products. The System/360 was not just a single computer but a family of machines, from an inexpensive model intended to replace the popular IBM 1401 to high-end computers that were optimized for numerical calculations. Because each model had the same instruction set (with exceptions), software written for a small model could be ported to a higher-end model as a customer's needs grew, thus preserving customers' investment in the programs they had developed.

With this announcement, IBM "bet the company," in the words of a famous magazine article of the day.[3] It invested enormous resources—unprecedented outside the federal government—in not only the new computers but also tape and disk drives, printers, card punches and readers, and a host of other supporting equipment. IBM also invested in a new operating system and a new programming language (Pl/1) that eventually were delivered but were less successful. Fortunately, extensions to existing FORTRAN and COBOL software, as well as operating systems developed independent of IBM's main line, saved the day. The announcement strained the company's resources, but by the end of the 1960s, IBM had won the bet. The company not only survived; it thrived—almost too much, as it was the target of a federal antitrust suit as a result of the increase in market share it had obtained by the late 1960s.

Did the System/360 in fact cover the full range of applications? Certainly there was no longer a need for customers to choose between a scientific-oriented machine (like the IBM 7090) or a comparable business data processing machine (the corresponding IBM product was the 7030). Low-end System/360 models did not extend downward into the PDP-8 minicomputer range. High-end models had trouble competing with so-called supercomputers developed by the Control Data Corporation, under the engineering leadership of the legendary designer Seymour Cray. In particular the Control Data CDC-6000, designed by Cray and announced around the time of IBM's System/360 announcement, was successfully marketed to U.S. laboratories for atomic energy, aerodynamics, and weather research. In 1972 Cray left Control Data to found a company, named after himself, that for the following two decades continued offering high-performance supercomputers that few competitors could match. The company did not worry much about the minicomputer threat, but it was concerned about Control Data's threat to the high end of the 360 line.

Another blow to the System/360 announcement was especially galling to IBM: researchers at MIT's Project MAC concluded that the System/360's architecture was ill suited for time-sharing. Although MIT had been using IBM 7090 mainframes for initial experiments, for the next phase of Project MAC, researchers chose a General Electric computer, which they thought would better suit their needs. This choice hardly affected IBM's sales—the company was delivering as many System/360s as it could manufacture. But some within IBM believed that the batch-oriented mode of operation and, by implication, the entire 360 architecture, was soon be obsolete in favor of conversational, time-shared systems. IBM responded by announcing the System/360, Model 67, which supported time-sharing, but its performance was disappointing. Eventually IBM was able to offer more robust time-sharing systems, but it was the networked personal workstation, not time-sharing, that overturned batch, and that did not come for many years later. MIT had trouble scaling up the time-sharing model, and we saw that the Advanced Research Projects Agency (ARPA), initially a strong supporter of time-sharing, turned to other avenues of research that it thought were more promising, including networking that led to the ARPANET (to which a number of System 360 computers were connected). Time-sharing did not go away; it evolved into client-server architecture, an innovation that came from a laboratory located in Silicon Valley and run by the Xerox Corporation. Large, dedicated mainframes, now called servers, store and manipulate massive amounts of data, delivering those data over high-speed networks to powerful personal computers, laptops, and other smart devices, which are nothing like the "dumb" terminals of the initial time-sharing model. These clients do a lot of the processing, especially graphics. The origins of client-server are intimately connected with the transition from the ARPANET to the Internet as it exists today, a story examined in more detail in chapter 6.

IBM was aware of the invention of the integrated circuit and its implications for computer design, but for the System/360 it chose instead to develop its own type of circuits, called solid logic technology, in which individual components were deposited on a ceramic wafer. The decision was based on IBM's ability to manufacture the devices reliably and in quantity, which was more important than the potential for the IC to surpass solid logic technology in performance. That was the case in 1964 when the announcement was made; however, the rapid pace of IC technology coming out of Silicon Valley caused IBM to reconsider that decision. It shifted to ICs for the follow-on System/370, introduced near the end of the decade. In most other respects, the System/370 was an evolutionary advance, keeping the basic architecture and product line of the 1964 announcement. By the mid-1970s, the mainframe world had made the transition to ICs, with IBM dominating, followed by a "BUNCH" of competitors: Burroughs, Univac, National Cash Register, Control Data, and Honeywell (Honeywell had bought General Electric's computer line in 1970, and UNIVAC, now Sperry-UINIVAC, took over RCA's customer base in 1971). This world coexisted somewhat peacefully with the fledgling minicomputer industry, dominated by Digital Equipment Corporation, whose main competitors were Data General, Hewlett-Packard, another division of Honeywell, Modular Computer Systems, and a few others.

5

THE MICROPROCESSOR

By 1970, engineers working with semiconductor electronics recognized that the number of components placed on an integrated circuit was doubling about every year. Other metrics of computing performance were also increasing, and at exponential, not linear, rates. C. Gordon Bell of the Digital Equipment Corporation recalled that many engineers working in the field used graph paper on which they plotted time along the bottom axis and the logarithm of performance, processor speeds, price, size, or some other variable on the vertical axis. By using a logarithmic and not a linear scale, advances in the technology would appear as a straight line, whose slope was an indication of the time it took to double performance or memory capacity or processor speed. By plotting the trends in IC technology, an engineer could predict the day when one could fabricate a silicon chip with the same number of active components (about 5,000) as there were vacuum tubes and diodes in the UNIVAC: the first commercial computer marketed in the United States in 1951.[1] A cartoon that appeared along with Gordon Moore's famous 1965 paper on the doubling time of chip density showed a sales clerk demonstrating dictionary-sized "Handy Home Computers" next to "Notions" and "Cosmetics" in a department store.[2]

But even as Moore recognized, that did not make it obvious how to build a computer on a chip or even whether such as device was practical. That brings us back to the Model T problem that Henry Ford faced with mass-produced automobiles. For Ford, mass production lowered the cost of the Model T, but customers had to accept the model as it was produced (including its color, black), since to demand otherwise would disrupt the mass manufacturing process.[3] Computer engineers called it the "commonality problem": as chip density increased, the functions it performed were more specialized, and the likelihood that a particular logic chip would find common use among a wide variety of customers got smaller and smaller.[4]

A second issue was related to the intrinsic design of a computer-on-a-chip. Since the time of the von Neumann Report of 1945, computer engineers have spent a lot of their time designing the architecture of a computer: the number and structure of its internal storage registers, how it performed arithmetic, and its fundamental instruction set, for example.[5] Minicomputer companies like Digital Equipment Corporation and Data General established themselves because their products' architecture was superior to that offered by IBM and the mainframe industry. If Intel or another semiconductor company produced a single-chip computer, it would strip away one reason for those companies' existence. IBM faced this issue as well: it was an IBM scientist who first used the term *architecture* as a description of the overall design of a computer. The success of its System/360 line came in a large part from IBM's application of microprogramming, an innovation in computer architecture that had previously been confined mainly to one-of-a-kind research computers.[6]

The chip manufacturers had already faced a simpler version of this argument: they produced chips that performed simple logic functions and gave decent performance, though that performance was inferior to custom-designed circuits that optimized each individual part. Custom-circuit designers could produce better circuits, but what was crucial was that the customer did not care. The reason was that the IC offered good enough performance at a dramatically lower price, it was more reliable, it was smaller, and it consumed far less power.[7]

It was in that context that Intel announced its 4004 chip, advertised in a 1971 trade journal as a "microprogrammable computer on a chip." Intel was not alone; other companies, including an electronics division of the aerospace company Rockwell and Texas Instruments, also announced similar products a little later. As with the invention of the integrated circuit, the invention of the microprocessor has been contested. Credit is usually given to Intel and its engineers Marcian ("Ted") Hoff, Stan Mazor, and Federico Faggin, with an important contribution from Masatoshi Shima, a representative from a Japanese calculator company that was to be the first customer. What made the 4004, and its immediate successors, the 8008 and 8080, successful was its ability to address the two objections mentioned above. Intel met the commonality objection by also introducing input–output and memory chips, which allowed customers to customize the 4004's functions to fit a wide variety of applications. Intel met the architect's objection by offering a system that was inexpensive and small and consumed little power. Intel also devoted resources to assist customers in adapting it to places where previously a custom-designed circuit was being used.

In recalling his role in the invention, Hoff described how impressed he had been by a small IBM transistorized computer called the 1620, intended for use by scientists. To save money, IBM stripped the computer's instruction set down to an absurdly low level, yet it worked well and its users liked it. The computer did not even have logic to perform simple addition; instead. whenever it encountered an "add" instruction, it went to a memory location and fetched the sum from a set of precomputed values.[8] That was the inspiration that led Hoff and his colleagues to design and fabricate the 4004 and its successors: an architecture that was just barely realizable in silicon, with good performance provided by coupling it with detailed instructions (called microcode) stored in read-only-memory (ROM) or random-access-memory (RAM) (see figure 5.1).

The Personal Computer

Second only to the airplane, the microprocessor was the greatest invention of the twentieth century. Like all other great inventions, it was disruptive as much as beneficial. The benefits of having all the functionality of a general-purpose computer on a small and rugged chip are well known. But not everyone saw it that way. The first, surprisingly, was Intel, where the microprocessor was invented. Intel marketed the invention to industrial customers and did not imagine that anyone would want to use it to build a computer for personal use. The company recognized that selling this device required a lot more assistance than was required for simpler devices. It designed development systems, which consisted of the microprocessor, some read-only and read-write memory chips, a power supply, some switches or a keypad to input numbers into the memory, and a few other circuits. It sold these kits to potential customers, hoping that they would use the kits to design embedded systems (say, for a process controller for a chemical

United States Patent [19]

[11] **3,821,715**

Hoff, Jr. et al.

[45] **June 28, 1974**

[54] **MEMORY SYSTEM FOR A MULTI-CHIP DIGITAL COMPUTER**

[75] Inventors: **Marcian Edward Hoff, Jr.**, Santa Clara; **Stanley Mazor**, Sunnyvale; **Federico Faggin**, Cupertino, all of Calif.

[73] Assignee: **Intel Corporation**, Santa Clara, Calif.

[22] Filed: **Jan. 22, 1973**

[21] Appl. No.: **325,511**

[52] **U.S. Cl.** .... **340/172.5**, 340/173 R, 340/173 SP, 307/238

[51] **Int. Cl.** **G06f 13/00**, G11c 11/44

[58] **Field of Search** .... 340/172.5, 173 SP, 173 R; 307/238, 279

[56] **References Cited**

UNITED STATES PATENTS

3,460,094	8/1969	Pryor	340/172.5
3,641,511	2/1972	Cricchi et al.	307/238 X
3,680,061	7/1972	Arbab et al.	340/173 R
3,681,763	8/1972	Meade et al.	340/173 R
3,685,020	8/1972	Meade	340/172.5
3,702,988	11/1972	Haney et al.	340/172.5
3,719,932	3/1973	Cappon	340/173 R
3,731,285	5/1973	Bell	340/172.5
3,735,368	5/1973	Beausoleil	340/173 R
3,737,866	6/1953	Gruner	340/172.5
3,740,723	6/1973	Beausoleil et al.	340/172.5

OTHER PUBLICATIONS

Schuenemann, "Computer Control" in IBM Technical Disclosure Bulletin, Vol. 14, No. 12, May 1972; pp. 3794–3795.

Primary Examiner—Paul J. Henon
Assistant Examiner—Melvin B. Chapnick
Attorney, Agent, or Firm—Spensley, Horn & Lubitz

[57] **ABSTRACT**

A general purpose digital computer which comprises a plurality of metal-oxide-semiconductor (MOS) chips. Random-access-memories (RAM) and read-only-memories (ROM) used as part of the computer are coupled to common bi-directional data buses to a central processing unit (CPU) with each memory including decoding circuitry to determine which of the plurality of memory chips is being addressed by the CPU. The computer is fabricated using chips mounted on standard 16 pin dual in-line packages allowing additional memory chips to be added to the computer.

17 Claims, 5 Drawing Figures

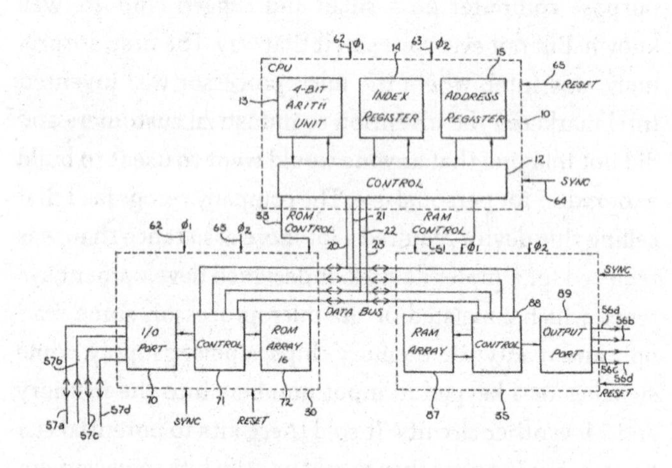

Figure 5.1 Patent for the microprocessor.

or pharmaceutical plant), and then place large orders to Intel for subsequent chips. These development kits were in effect genuine computers and a few recognized that, but they were not marketed as such.

Hobbyists, ham radio operators, and others who were only marginally connected to the semiconductor industry thought otherwise. A few experimental kits were described in electronics hobby magazines, including *73* (for ham radio operators) and *Radio Electronics*. Ed Roberts, the head of an Albuquerque, New Mexico, supplier of electronics for amateur rocket enthusiasts, went a step further: he designed a computer that replicated the size and shape, and had nearly the same functionality, as one of the most popular minicomputers of the day, the Data General Nova, at a fraction of the cost. When Micro Instrumentation and Telemetry Systems (MITS) announced its "Altair" kit on the cover of the January 1975 issue of *Popular Electronics*, the floodgates opened (see figure 5.2).

The next two groups to be blindsided by the microprocessor were in the Boston region. The impact on the minicomputer companies of having the architecture of, say, a Data General minicomputer on a chip has already been mentioned. A second impact was in the development of advanced software. In 1975, Project MAC was well underway on the MIT campus with a multifaceted approach toward innovative ways of using mainframe computers interactively. A lot of their work centered around developing

Just as the availability of fast and reliable transistors after 1965 fueled a number of mini-computer companies, so too did the availability of microprocessors and associated memory chips fuel a personal computer firestorm.

Figure 5.2 The Altair personal computer. Credit: Smithsonian Institution.

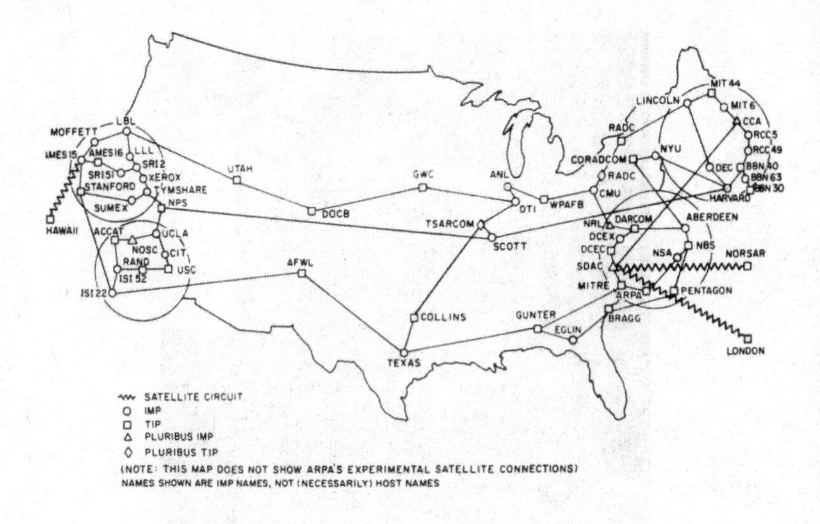

Figure 5.3 Map of the ARPANET, circa 1974. Notice the concentration in four regions: Boston, Washington, D.C., Silicon Valley, and southern California. (Credit: DARPA)

Figure 5.4 Historical marker, Wilson Boulevard, Arlington, Virginia, commemorating the location of ARPA offices, where the specifications for the ARPANET were developed in the early 1970s. The common perception of the Internet is that it resides in an ethereal space; in fact, its management and governance is located in the Washington, DC, region, not far from the Pentagon. Photo by the author.

incremental improvements. Others used microprocessors offered by competitors to Intel. Among the most favored was a chip called the 6502, sold by MOS Technology (the term *MOS* referred to the type of circuit, metal-oxide-semiconductor, which lends itself to a high density of integration on a chip).[9] The 6502 was introduced at a 1975 trade show and was being sold for only $25.00. Other personal systems used chips from Motorola and Zilog, whose Z-80 chip extended the instruction set of the Intel 8080 used in the Altair. Just as Intel built development systems to allow potential customers to become familiar with the microprocessor's capabilities, so too did MOS Technology offer a single board system designed around the 6502. The KIM-1 was primitive, yet to the company's surprise, it sold well to hobbyists and enthusiasts who were eager to learn more about this phenomenon.[10] Radio Shack chose the Zilog chip for a computer, the TRS-80, which it sold in its stores, and a young Silicon Valley enthusiast named Steve Wozniak chose the 6502 for a system he was designing, later called the Apple 1.

The transformation of the computer from a room-sized ensemble of machinery to a handheld personal device is a paradox. On the one hand, it was the direct result of advances in solid-state electronics following the invention of the transistor in 1947 and, as such, an illustration of technological determinism: the driving of social change by technology. On the other hand, personal computing was unanticipated by computer engineers; instead, it was driven by idealistic visions of the 1960s-era counterculture. By that measure, personal computing was the antithesis of technological determinism. Both are correct. In Silicon Valley, young computer enthusiasts, many of them children of engineers working at local electronics or defense firms, formed the Homebrew Computer Club and met regularly to share ideas and individual designs for personal computers. The most famous of those computers was the Apple. It was the best known, and probably the best designed, but it was one of literally dozens of competing microprocessor-based systems from all over the United States and in Europe. Other legendary personal systems that used the 6502 were the Commodore PET; its follow-on, the Commodore 64, used an advanced version of the chip and was one of the most popular of the first-generation personal computers. The chip was also found in video game consoles—special-purpose computers that enjoyed a huge wave of popularity in the early 1980s.

MITS struggled, but Apple, founded in 1976, did well under the leadership of Wozniak and his friend, Steve Jobs (a third founder, Ronald Gerald Wayne, soon sold his share of the company for a few thousand dollars, fearing financial liability if the company went under).[11] The combination of Wozniak's technical talents and Jobs's vision of the limitless potential of personal computing set that company apart from the others. The company's second consumer product, the Apple II, not only had an elegant design; it also was attractively packaged and professionally marketed. Apple's financial success alerted those in Silicon Valley that microprocessors were suitable for more than just embedded or industrial uses. It is no coincidence that these computers could also play games, similar to what was offered by the consoles from companies like Atari. That was a function that was simply unavailable on a mainframe, for economic and social reasons.

Bill Gates and Paul Allen recognized immediately that the personal computer's potential depended on the availability of software. Their first product for the Altair was a program, written in 8080 machine language, that allowed users to write programs in BASIC, a simple language developed at Dartmouth College for undergraduates to learn how to program. At Dartmouth, BASIC was implemented on a General Electric mainframe that was time-shared; the BASIC that Gates and Allen delivered for the Altair was to be used by the computer's owner, with no sharing at all. That was a crucial distinction. Advocates of large time-shared or networked systems did not fully grasp why people wanted a computer of their own, a desire that drove the growth of computing for the following decades. The metaphor of a computer utility, inspired by electric power utilities, was seductive. When such a utility did come into existence, in the form of the World Wide Web carried over the Internet, its form differed from the 1960s vision in significant ways, both technical and social.

Other ARPA-funded researchers at MIT, Stanford, and elsewhere were working on artificial intelligence: the design of computer systems that understood natural language, recognized patterns, and drew logical inferences from masses of data. AI research was conducted on large systems, like the Digital Equipment Corporation PDP-10, and it typically used advanced specialized programming languages such as LISP. AI researchers believed that personal computers like the Altair or Apple II were ill suited for those applications. The Route 128 companies were making large profits on minicomputers, and they regarded the personal computers as too weak to threaten their product line. Neither group saw how quickly the ever-increasing computer power described by Moore's law, coupled with the enthusiasm and fanaticism of hobbyists, would find a way around the deficiencies of the early personal computers.

While the Homebrew Computer Club's ties to the counterculture of the San Francisco Bay Area became legendary, more traditional computer engineers were also recognizing the value of the microprocessor by the late 1970s.[12] IBM's mainframes were profitable, and the company was being challenged in the federal courts for antitrust violations. A small IBM group based in Florida, far from IBM's New York headquarters, set out to develop a personal computer based around an Intel microprocessor. The IBM

Personal Computer, or PC, was announced in 1981. It did not have much memory capacity, but in part because of the IBM name, it was a runaway success, penetrating the business world. Its operating system was supplied by Microsoft, which made a shrewd deal allowing it to market the operating system (known as MS-DOS) to other manufacturers. The deal catapulted Microsoft into the ranks of the industry's wealthiest as other vendors brought out machines that were software-compatible clones of the IBM PC. Among the most successful were Compaq and Dell, although for a while, the market was flooded with competitors. In the 1970s, IBM had fought off a similar issue with competitors selling mainframe components that were, as they called them, "plug-compatible" with IBM's line of products (so-called because a customer could unplug an IBM component such as a disk drive, plug in the competitor's product, and the system would continue to work). This time IBM was less able to assert control over the clone manufacturers. Beside the deal with Microsoft, there was the obvious factor that most of the innards of the IBM PC, including its Intel microprocessor, were supplied by other companies. Had IBM not been in the midst of fighting off an antitrust suit, brought on in part by its fight with the plug-compatible manufacturers, it might have forestalled both Microsoft and the clone suppliers. It did not, with long-term implications for IBM, Microsoft, and the entire personal computer industry.

One reason for the success of both the IBM PC and the Apple II was a wide range of applications software that enabled owners of these systems to do things that were cumbersome to do on a mainframe. These first spreadsheet programs—VisiCalc for the Apple and Lotus 1-2-3 for the IBM PC—came from suppliers in Cambridge, Massachusetts; they were followed by software suppliers from all over the United States. Because of their interactivity, spreadsheet programs gave more direct and better control over simple financial and related data than even the best mainframes could give. From its base as the supplier of the operating system, Microsoft soon dominated the applications software business as well. And like IBM, its dominance triggered antitrust actions against it from the U.S. Department of Justice. A second major application was word processing—again something not feasible on an expensive mainframe, even if time-shared, given the massive, uppercase-only printers typically used in a mainframe installation. For a time, word processing was offered by companies such as Wang or Lanier with specialized equipment tailored for the office environment. These were carefully designed to be operated by secretaries (typically women) or others with limited computer skills. The vendors deliberately avoided calling them "computers," even though that was exactly what they were. As the line of IBM personal computers with high-quality printers matured and as reliable and easy-to-learn word processing software became available, this niche disappeared.

Xerox PARC

Apple's success dominated the news about Silicon Valley in the late 1970s, but equally profound innovations were happening in the valley out of the limelight, at a laboratory set up by the Xerox Corporation. The Xerox Palo Alto Research Center (PARC) opened in 1970, at a time when dissent against the U.S. involvement in the war in Vietnam ended the freewheeling funding for computer research at the Defense Department. An amendment to a defense authorization bill in 1970, called the Mansfield amendment named after Senator Mike Mansfield who chaired the authorization committee, contained language that restricted defense funding to research that had a direct application to military needs. The members of that committee assumed that the National Science Foundation (NSF) would take over some of ARPA's more advanced research, but that did not happen (although the NSF did take the lead in developing computer networking). The principal beneficiary of this legislation was Xerox: its Palo Alto lab was able to hire many of the top scientists previously funded by ARPA, and from that laboratory emerged innovations that defined the digital world: the Windows metaphor, the local area network, the laser printer, the seamless integration of graphics and text on a screen. Xerox engineers developed a computer interface that used icons—symbols of actions—that the user selected with a mouse. The mouse had been invented nearby at Stanford University by Douglas Engelbart, who was inspired to work on human-computer interaction after reading Vannevar Bush's "As We May Think" article in the *Atlantic*.[13] Some of these ideas of using what we now call a graphical user interface had been developed earlier at the RAND Corporation in southern California, but it was at Xerox PARC where all these ideas coalesced into a system, based around a computer of Xerox's design, completed in 1983 and called the Alto.

Xerox was unable to translate those innovations into successful products, but an agreement with Steve Jobs, who visited the lab at the time, led to the interface's finding its way to consumers through the Apple Macintosh, introduced in 1984. The Macintosh was followed by products from Microsoft that also were derived from Xerox, carried to Microsoft primarily by Charles Simonyi, a Xerox employee whom Microsoft hired. Xerox innovations have become so common that we take them for granted, especially the ability of word processors to duplicate on the screen what the printed page will look like: the so-called WYSIWYG, or "what you see is what you get," feature (a phrase taken from a mostly forgotten television comedy *Laugh In*).

Consumers are familiar with these innovations as they see them every day. Even more significant to computing was the invention, by Xerox researchers Robert Metcalfe and David Boggs, of Ethernet.[14] This was a method of connecting computers and workstations at the local level, such as an office or department on a college campus. By a careful mathematical analysis, in some ways similar to the notion of packet switching for the ARPANET, Ethernet allowed data transfers that were orders of magnitude faster than what had previously been available. Ethernet finally made practical what time-sharing had struggled for so many years to implement: a way of sharing computer resources, data, and programs without sacrificing performance. The technical design of Ethernet dovetailed with the advent of microprocessor-based personal computers and workstations to effect the change. No longer did the terminals have to be "dumb," as they were marketed; now the terminals themselves had a lot of capability of their own, especially in their use of graphics.

6

THE INTERNET AND THE WORLD WIDE WEB

As these events were unfolding in Silicon Valley, the ARPANET continued expanding. The nineteen nodes connected in 1971 doubled within two years. By 1977, the year the Apple II was introduced, there were about sixty nodes. (A "node" could be a computer connected to the network or a simple terminal, such as the Teletype Model ASR-33.)[1] The Stanford laboratory where Douglas Engelbart invented the mouse was one of the first to have an ARPANET connection by 1970, and the Xerox Palo Alto Research Lab had a connection by 1979. In 1983 ARPANET switched from the original protocol for routing packets to a new set that was better able to handle the growing network. The original protocol was divided into two parts: the Transmission Control Protocol, to manage the assembly of packets into messages and ensure that the original message was received, and the Internet Protocol to pass packets from one node to another.[2] The combination, TCP/IP, was primarily the work of two scientists, Vint Cerf and Robert Kahn, and it remains at the basis of the Internet to this day.

This decision to split the protocol was based in part on a need to ensure robustness in a military environment. The initial ARPANET was designed to ensure reliable communications during a time of international crisis or even nuclear war—recall the November 1962 conference that J. C. R. Licklider attended right after the Cuban missile crisis. One of the originators of the concept of packet switching, Paul Baran of the RAND Corporation, stated the need to communicate during a crisis explicitly as a goal. The modern Internet evolved from those initial decisions made when the ARPANET was being built. The Internet is much more than that, of course. It is a social and political as well as technical entity. It relies on TCP/IP, but nearly everything else about it is different.

ARPANET nodes were connected by communication lines leased from AT&T or other commercial suppliers at speeds of about 50,000 bits per second. That gave way to microwave radio and satellites. Traffic is now carried mainly on fiber-optic lines with orders of magnitude greater speed. Likewise, the mainframes and minicomputers of the ARPANET evolved to massive server farms consisting of large arrays of microprocessor-based systems similar to desktop personal computers. The mesh topology of the early network evolved to a set of high-speed "backbones" crossing the globe, with a hierarchy of subnetworks eventually connecting personal devices at the individual or office level. Although the Internet is truly a global phenomenon, the east-to-west backbones across the United States dominate traffic. The topology of this network dovetailed with an innovation not foreseen by the ARPANET pioneers: local area networks, especially Ethernet. Outside the home, users access the Internet through a desktop computer or workstation connected to Ethernet, which is connected to a regional network and then to a backbone.

The governance of the Internet has also evolved. In 1983 the Department of Defense established a network for its internal use (MILNET). It turned over the management of the rest of the network to the National Science Foundation (NSF), which in turn let contracts to commercial suppliers to build a network for research and academic use. By this time, commercial networks using packet switching began to appear. Within a few years the "Internet," as it was now being called, was fully open to commercial use.[3] Since about 1995, the U.S. Department of Commerce has held authority for Internet governance, including the assignment of names, although Commerce waited until 2005 to state this formally.[4]

The main difference between the ARPANET and the Internet is the social and cultural component. Once again the forces that drove the personal computer phenomenon, coexisting alongside the ARPA-funded work, come into play.

Early adopters of PCs used them for games, later spreadsheets and word processing, but not for communicating. Nevertheless the PC had the potential to be networked. Like everything else about PCs, networking required effort by PC owners. Networking was not as well publicized as other applications, but evidence of its significance can be found in some of the publications that appeared in the early days of personal computing. Beginning in 1977, the first of many editions of *The Complete Handbook of Personal Computer Communications,* by Alfred Glossbrenner, appeared. The author argued that those who were not taking advantage of the many online services then available were missing the most important dimension of all of the personal computer revolution.[5] He listed dozens of databases that users could access over their home telephone lines. These databases began as specialized services for corporate or professional customers and were now being made available to individuals to amortize the costs of establishing them and broaden their market. It was a positive feedback cycle: as more people accessed these services, the providers saw a reason to offer more types of information, which in turn spurred more customers. Glossbrenner was not alone in his evangelism. Alan Kay, of the Xerox Palo Alto Research Center, remarked that "a computer is a communications device first, second, and third." And Stewart Brand of the *Whole Earth Catalog*, stated, "`Telecommunicating' is our founding domain."[6] In 1977 however, the PC was not first and foremost a communications device. It was just barely a computer at all. Just as it took visionaries like Steve Jobs and Steve Wozniak to make the PC a viable machine, it also took visionaries like Brand, Glossbrenner, and Kay to make telecommunicating for those outside the privileged world of the ARPANET a reality.

To connect a PC, an individual connected a modem, which translated computer signals into audio tones that the telephone network recognized and vice versa. The connection to the phone network was made by placing the handset into an acoustic cradle; later this was replaced by a direct wire. The user then dialed a local number and connected to a service. In addition to commercial services, one could dial into a personal local "bulletin board," which may have been a simple personal computer, augmented by the addition of a hard drive memory unit and run by a neighbor. If the service was not located in the local calling zone, the user would be reluctant to dial a long-distance number given the rate structure of long-distance calls in those days of the AT&T monopoly. Bulletin boards got around this restriction by saving messages or files and forwarding them to distant bulletin boards late at night, when calling rates were lower. Commercial services took another route: they established local phone numbers in metropolitan areas and connected the lines to one another over their own private network. AT&T made no distinction between the sounds of a voice and the staccato tones of data sent over the phone, so a local data connection for an individual appeared to be "free." The services established different pricing schemes. For example CompuServe charged $19.95 for an initial setup and one hour of connection to its service, plus an additional $15.00 an hour during business hours, dropping to $12.50 an hour on evenings and weekends. The Source charged a $10.00 monthly fee, plus a fee per hour of connect time depending on the speed, and so on.[7] NEXIS charged $50.00 a month, plus $9.00 to $18.00 per search during business hours (this service was obviously not intended for casual, personal use).[8] Most of these services maintained a central database on mainframe computers.[9] The topology of these services was that of a hub and spokes, harking back to the time-sharing services from which they descended.

Years before the term *cyberspace* became fashionable, Glossbrenner recognized that the Source, operating out of northern Virginia, was creating a new realm of social interaction, one that Licklider envisioned only among scientists and the military. Every day, as millions of people look at and update their Facebook accounts, the Source lives on. After its demise in 1989, Glossbrenner noted, "It was the system that started it all. It made the mistakes and experienced the successes that paved the way for those who followed. Above all, it contributed a *vision* of what an online utility could and should be."[10]

The Source was the brainchild of William von Meister, whose ideas ran ahead of an ability to implement them.

In 1979, von Meister was pushed out of the service he founded.[11] Undeterred, he founded another service, to deliver songs to consumers over the phone, and then another company, Control Video Corporation, whose intent was to allow owners of Atari game consoles to play interactively with others online.[12] The service attracted the notice of Steve Case, a marketer from Pizza Hut. Case was converted after logging on from a Kaypro computer from his apartment in Wichita, Kansas. He joined Control Video and tried to revive it. In 1985, with funding from a local Washington entrepreneur, he formed Quantum Computer Services, later renamed America Online (AOL).[13] By the late 1990s it connected more individuals—as many as 30 million—to the Internet than any other service.

AOL was not alone. In 1984 Prodigy was founded, with joint support from CBS, IBM, and Sears. It pioneered the use of computer graphics at a time when PCs ran the text-only DOS operating system and when modems had very low speeds. Prodigy preloaded graphics onto the user's computer to get around those barriers. The graphics not only made the service more attractive; they provided advertisers a way to place ads before subscribers, and those ads offset most of the costs. Thus, Prodigy anticipated the business model of the World Wide Web. Prodigy suffered from its users' heavy use of e-mail and discussion forums, which used a lot of connect time but did not generate ad views. The service also attempted to censor the discussions, which led to a hostile reaction from subscribers. America Online had a more freewheeling policy in its chat rooms, where it appointed volunteers to monitor the content of the rooms. These volunteers guided chat room discussions with a very light touch of censorship. Many of the chat rooms focused on dating, meeting partners, or flirting (helped by the fact that one's physical appearance did not factor into the discussions, and also that one could disguise one's gender or sexual preference while chatting). Like Facebook twenty years later, a major reason for AOL's success was hidden in plain sight.

AOL's business model, if it was at all understood by the NSF, would not have been something the NSF would have advertised. One must be constantly reminded that these efforts by Glossbrenner, Brand, and the others were operating independent of the NSF-sponsored Internet. For example, Stewart Brand's path-breaking *Whole Earth Software Catalog*, published in 1984, did not mention the Internet at all.[14] Neither did Glossbrenner in the early editions of his book, and in later editions he mentions "the internet" (lowercase) as one of many networks that were in existence.

Halfway between the Internet and these personal connections were several networks that also contributed to the final social mix of cyberspace. One was BITNET, built in 1981 to connect IBM mainframes.[15] The service was often the first entree to networking for many liberal arts students and faculty, who otherwise had no access.[16] One of its features was the "Listserv": a discussion forum similar to those provided by the Source and CompuServe and still in use today. Another service, Usenet, began the same way in 1980, as an interconnection between computers that used the Unix operating system. This was an operating system developed at Bell Laboratories, which had an enthusiastic base of programmers among minicomputer users. The name was a subtle jab at the operating system Multics, which had been developed for Project MAC and had experienced problems getting into production. Usenet was not at first connected to the Internet but had a large impact nonetheless. Thanks to ARPA funding, the TCI/IP protocols were bundled into a distribution of Unix, which became a standard for Internet software by the mid-1980s. Eventually personal computers using the Microsoft Windows operating system could also connect to the Internet, but people writing networking software had to be fluent in Unix, and they spent a lot of time discussing programming and related technical topics on Usenet groups. Usenet groups tended to be more free-wheeling than those on BITNET, reflecting the subculture of Unix programmers.

The Role of the National Science Foundation

The NSF became involved in networking because it wanted to provide access to supercomputers: expensive devices that were to be installed in a few places with NSF support. By 1986 the NSF linked five supercomputer centers, and it made three crucial decisions that would affect later history. The first was to adopt the TCP/IP protocols promulgated by ARPA; although this seems obvious in hindsight, it was not at the time.[17] The second was to create a general-purpose network, available to researchers in general, not just to a specific discipline. The third was to fund the construction of a high-speed backbone, to which not only these centers but also local and regional networks would connect.[18] In 1987 the NSF awarded a contract to replace the original backbone with a new one, at a speed known as T1: 1.5 million bits per second (Mbps). In 1992 the NSFNET was upgraded to 45 Mbps—T3. (Compare these speeds to the few hundreds of bits per second available to home computer users with modems connected to their telephones.)

One of the major recipients of these contracts was MCI, which to this day (now as a subsidiary of Verizon) is the principal carrier of Internet backbone traffic. MCI descended from a company founded in 1963 to provide a private microwave link between Chicago and St. Louis. AT&T fought MCI at every step, and only in 1971 was MCI able to offer this service. Having fought that battle and won, MCI changed course. The company saw that voice traffic was becoming a low-profit commodity, while packet-switched data traffic was growing exponentially, so it began to focus on providing backbone service to data networks. Contractors delivered a T1 backbone to the NSF by mid-1988. By 1990 the net connected about 200 universities, as well as other government networks, including those operated by NASA and the Department of Energy. It was growing rapidly. BITNET and Usenet established connections, as did many other domestic and several international networks. With these connections in place, the original ARPANET became obsolete and was decommissioned in 1990.

Commercial networks were allowed to connect to the backbone, but that traffic was constrained by what the NSF called an "acceptable use policy." The Congress could not allow the NSF to support a network that others were using to make profits from. For-profit companies could connect to the network but not use it for commercial purposes.[19] In 1988 the NSF allowed MCI to connect its MCI e-mail service, initially for "research purposes"—to explore the feasibility of connecting a commercial mail service.[20] That decision was probably helped by Vint Cerf, who had worked on developing MCI Mail while he was employed there. The MCI connection gave its customers access to the growing Internet, and not long after, CompuServe and Sprint got a similar connection. Under the rubric of "research," a lot could be accomplished.

The topology of the NSF network encouraged regional networks to connect to the NSF's backbone. That allowed entrepreneurs to establish private networks and begin selling services, hoping to use a connection to the backbone to gain a national and eventual global reach. The NSF hoped that the acceptable use policy would encourage commercial entities to fund the creation of other backbones and allow the NSF to focus on its mission of funding scientific research. By 1995 all Internet backbone services were operated by commercial entities. By then the acceptable use policy was relaxed. That milestone was contained in an amendment to the Scientific and Technology Act of 1992, which pertained to the authorization of the NSF. The legislation passed and was signed into law by President George H. W. Bush on November 23, 1992.[21] Paragraph g reads, in full:

> In carrying out subsection (a) (4) of this section, the Foundation is authorized to foster and support access by the research and education communities to computer networks which may be used substantially for purposes *in addition to* research and education in the sciences and engineering, if the additional uses will tend to increase the overall capabilities of the networks to support such research and education activities [emphasis added].[22]

With those three words, "in addition to," the modern Internet was born, and the NSF's role in it receded.[23]

The World Wide Web

For the layperson, the Internet is synonymous with a program, introduced in 1991, that runs on it: the World Wide Web. The progression of networking helps explain the confusion. The ARPANET was led by military people who wondered at times whether even personal e-mail messages would be permitted over it. Commercial airline pilots and air traffic controllers, for example, are forbidden to use their radios for anything except the business of flying the aircraft when landing or taking-off. That was the model that ARPANET funders had in mind. Companies like IBM and DEC had their own proprietary networks, which were used mainly for commercial purposes. Hobbyists who ran networks from their homes stressed the same idealism that drove the personal computer. The Internet subsumed all of those models, in part by virtue of its open protocols, lack of proprietary standards, and ability to interconnect (hence the name) existing networks of various designs. The Web, developed by Tim Berners-Lee and Robert Cailliau at the European Council for Nuclear Research (CERN) near Geneva, continued this trend by allowing the sharing of diverse kinds of information seamlessly over the Internet.[24] The software was, and remains, free.

Berners-Lee described his inspiration as coming from the diverse groups of physicists at CERN who would meet randomly at strategic places in the hallways of the center and exchange information among one another. Computer networks had places for a structured exchange of information, but it lacked that serendipity that no one can force but Berners-Lee felt he could facilitate. He was aware of a concept known as hypertext, or nonlinear writing, that Vannevar Bush had hinted at in his 1945 *Atlantic* essay, and which Apple had implemented as a stand-alone program, Hypercard, for its Macintosh line of computers. Doug Engelbart had developed a system that used such texts, the selection of which used his invention, the mouse. Hypertext had also been promoted by Ted Nelson, a neighbor of Stewart Brand in northern California and author of the self-published manifesto *Computer Lib.* As he saw the growing amounts of data on the Internet, Berners-Lee explored a way of porting those concepts to the online world. His invention had three basic components. The first was a uniform resource locator (URL), which took a computer to any location on the Internet—across the globe, down the hall, or even on one's own hard drive—with equal facility—a "flat" access device. The second was a protocol, called hypertext transfer protocol (http), which rode on top of the Internet protocols and facilitated the exchange of files from a variety of sources, regardless of the machines they were residing on. The third was a simple hypertext markup language (HTML), a subset of a formatting language already in use on IBM mainframes. That HTML was simple to learn made it easy for novices to build Web pages. With administrative and technical support from Robert Cailliau, the system was working on machines at CERN on Christmas Day 1990, and the following year it began to spread around the globe.[25]

Berners-Lee also wrote a program, called a browser, that users installed on their computer to decode and properly display the information transmitted over the Web (the term may have come from Apple's Hypercard). In 1992, as the Web began spreading across university and government laboratories in the United States, a group at an NSF-sponsored supercomputer center in Illinois took the browser concept further, adding rich graphics capabilities and integrating it seamlessly with the mouse and with icons. That browser was called Mosaic; its creators later moved to Silicon Valley and founded a company, Netscape, to market a similar commercial version of it. Microsoft later would license another descendant of Mosaic for its own browser, Internet Explorer. One significant feature of Netscape's browser that was not present in others was a way to transmit information, especially credit card numbers, securely by encrypting the data stream. The method, called Secure Socket Layer, opened up the Internet to commercial transactions. That phenomenon was soon dominated by firms like Amazon and eBay, both launched in 1995. Another feature that Netscape introduced was a way of tracking a user's interaction with a Web site over successive screens, an obvious feature but something not designed into the Web architecture. Netscape's browser did so by an identifying number called a "cookie" (the name no doubt came from the television program *Sesame Street*). When Netscape applied for an initial public offering of stock in August 1995, the resulting frenzy set off a bubble that made all the other Silicon Valley doings pale by comparison. The bubble eventually burst, but in its wake, the Web had established itself as not only an information service but a viable method of doing business.

By 1995 many American households were gaining access to the Internet. Initially the process was difficult. It required loading communications and other software on one's computer, establishing an account with an Internet service provider, and accessing the network by dialing a number on one's voice telephone line. As the decade progressed, the process became easier, with much of the software loaded onto a personal computer at purchase. In most urban areas, dial-up telephone connections were replaced by faster digital subscriber line (DSL) service, or one's cable television wire. Toward the end of the 1990s, a new method of access appeared, called Wi-Fi, which began as an ad hoc system that took advantage of little-used portions of the electromagnetic spectrum. Later it was sanctioned by the Institute of Electrical and Electronics Engineers (IEEE), which published specifications for it as Standard 802.11b. In 2003 Wi-Fi got a further boost from Intel, which developed circuits called "Centrino," with embedded Wi-Fi capability. Laptop computers, by then replacing desktop PCs for many users, would thus have wireless access built in. By design, Wi-Fi had a limited range. Many coffee shops, restaurants, airports, and other public spaces installed the service, either for a fee or free.[26]

Although Wi-Fi was popular, efforts to install it throughout metropolitan areas faltered, as did efforts to develop a faster and more powerful successor to 802.11b. There were several reasons for that, among them the rise of cellular phone networks that allowed users access to e-mail and other limited Internet services over their cell phone frequencies.

The Smart Phone

The rise of cell phone text communication, led by the Canadian company Research in Motion and its phone, the BlackBerry, is another case of history's stubborn refusal to follow a logical, linear path. The inventors of cellular telephone service knew they were advocating a revolutionary change in the way telephones connected to one another, but they did not foresee the impact their invention would have on computing. Mobile telephone service had been available for personal automobiles as far back as the 1950s, but with severe limitations. A phone call would use a single-frequency channel, of which few were allocated for such use, and the radios had to be powerful to cover a reasonable geographical area that the automobile might traverse. Just as packet switching replaced the dedicated circuits of traditional telephone connections, so too did cellular service break up a region into small areas, called cells, with the phones in each cell using frequencies that could be used without interference by other phones in other cells, since they operated at lower power. The technique worked by transferring, or handing off, a call from one cell to another as the caller moved. This technique obviously required a complex database of phone subscribers, a way to measure the relative strengths of their signals relative to nearby towers, and a switching method to hand over the call from one cell to another as the neighboring cell came into play.

The theory for such a service was developed at Bell Labs after World War II, but it was not practically implemented until years later. One of the pioneering installations came in 1969 on Amtrak's Northeast Corridor, between Washington, D.C., and New York, where passengers on board a train could make calls that were handed off from one tower to another as the train traveled down the track. Further details of this history are beyond the scope of this narrative, but it is widely agreed that the "first" call made on a handheld phone in the United States was by Martin Cooper, of Motorola, on April 3, 1973, to a scientist at Bell Labs.[27] Motorola's aggressive promotion of the service led to its dominance of the market for many years.

The relationship between these stories and the evolution of computing is similar to that of the Morse telegraph and the Teletype to computing's early history: continuous convergence of different base technologies. The Internet has been described as a convergence of the telegraph and traditional mainframe computer. Beginning around 2000, there was a second convergence, this time of personal computing with the radio and the telephone.

The BlackBerry allowed access to one's corporate e-mail account and to send and receive brief e-mail messages. It soon became addictive for executives. Another trigger for the convergence of computer and telephone came from a Silicon Valley company that had more modest aspirations. In 1996 Palm, Inc. introduced a personal organizer, soon dubbed a "personal digital assistant," that replaced a paper address book, to-do list, calendar and notebook that people carried in their pockets. The Palm Pilot was not the first of these devices, but thanks to a careful design by one of its founders, Jeff Hawkins, it had a simple, easy-to-understand user interface that made it instantly popular. The Palm's user-interface designers, as well as Martin Cooper of cell phone fame, were both influenced by the controls and especially the "Communicator" that Captain Kirk used in the original *Star Trek* television series.[28] Captain Kirk's Communicator had a flip-open case, which was popular among cell phones for a while (it has since fallen out of favor).

Not long after Palm established itself, Hawkins and two other employees from the company broke off and founded a competitor, Handspring, that offered similar devices. In 2002 Handspring offered a PDA with cell phone capabilities (or perhaps it was the other way around—a cell phone with PDA capabilities). Just as the Pilot was not the first digital organizer, the Handspring Treo was not the first smart phone. But it integrated the functions of phone and computer better than any other products, and it set off the next wave of convergence in computing. The convergence was triggered not only by advances in semiconductor and cell phone technology but also by a focus on the user's interaction with the device—a theme that goes back to World War II. Before long, smart phones were introduced with Wi-Fi as well as cellular reception, global positioning system (GPS) navigation, cameras, video, even gyroscopes and accelerometers that used tiny versions of components found in missile guidance systems (see figure 6.1).

Handspring and Palm later merged but were unable to hold on to their lead. In 2007 Apple introduced its iPhone, setting off a frenzy similar to the dot-com bubble of the 1990s that continues to this day. Apple followed the iPhone a few years later with the iPad, a tablet computer that also operated on cell phone as well as Wi-Fi networks. Computer companies had been offering tablet computers for years, but the iPad has been far more successful. Zeno's paradox prevents telling this story further—anything I

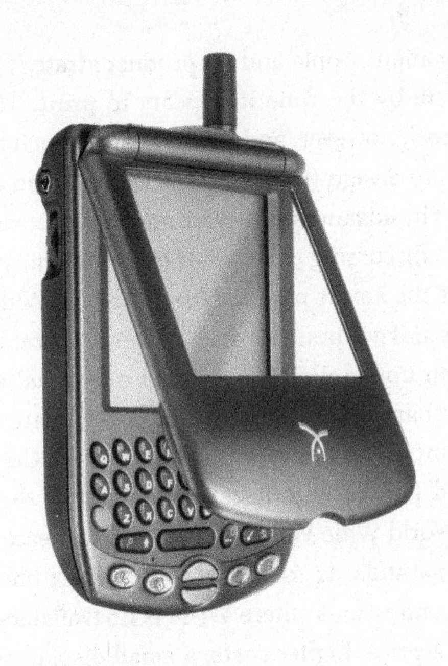

Figure 6.1 The Handspring Treo. The Treo inaugurated the class of devices known in the United States as the smart phone: a convergence of a host of digital technologies in a handheld portable device. Credit: Copyright © Hewlett-Packard Development Company, L.P. Reproduced with permission.

could say about Apple and its product strategy will likely be obsolete by the time it appears in print. The convergence is real, however, and it is driven as much by Apple's attention to design (following the lead of Palm and Handspring) as by advances in semiconductor electronics.

When discussing this convergence, one must acknowledge that the smart phone, like the World Wide Web before it, is not without its flaws. These portable devices operate on both cellular and Wi-Fi networks, which are not interchangeable. Thus the devices violate the basic goals of interoperability espoused by Vint Cerf, Robert Kahn, and Tim Berners-Lee, the pioneers of the Internet and the World Wide Web, who worked to create a system with open standards accessible to all. Smart phones offer wireless connections where Wi-Fi is unavailable. But that comes at a price: higher costs, a small display, no mouse, an awkward method of entering data, and (in the United States), a lock-in to a specific provider that takes steps to prevent migration to a competitor.

This evolution of the smart phone illustrates the significance of the microprocessor in the history of technology. An iPhone or iPad is a device built around a microprocessor. It accepts strings of bits as its input and sends out stream of bits as its output. Programmers determine the relationship between the two streams. Incoming bits include digitized radio signals, sounds, pixels from a camera, readings from a GPS receiver or accelerometer, text or equations typed on a touch screen—"anything" in the sense of a Universal Turing Machine. The output stream of bits can be voice, pictures, movies, sounds, text, mathematical symbols—again, anything.

Social Networking and Google

Smart phones addressed one deficiency of the Internet's connectivity. The initial Web design had several other deficiencies, which were addressed by a variety of approaches between its debut in 1991 and the turn of the millennium. Netscape's introduction of "cookies" and the Secure Socket Layer were among the first. Tim Berners-Lee described how he hoped that it would be as easy to write to the Web as it was to read pages on it. Although the simplicity of HTML allowed a computer-savvy person to create a Web site, it was not as easy as surfing through the Web with a mouse. By the late 1990s, that imbalance was addressed by a specialized Web site called a "Web log," soon shortened to the unfortunate term *blog*. Online postings appeared from the earliest days of local bulletin board systems (BBS). CompuServe offered something similar, which it compared to the freewheeling discussions on then-popular citizens band radio. Programs that automated or simplified this process for the Web began to appear around 1997, with blogger.com one of the

With the world literally at one's fingertips (or mouse click), how does one locate information? Unlike many of the proprietary, closed networks that preceded it, the Web came with no indexing scheme.

it. These evolved to be financially supported by advertisements, and during the inflation of the dot-com bubble in the late 1990s, dozens of such sites were founded. Most failed during the stock market crash of 2000–2002. A few deserve mention, especially Yahoo! and Google, which became among the most visited sites on the Web.

Before the World Wide Web prevailed on the Internet, AOL had pioneered the notion of a network that guided its subscribers to content in a friendly way. AOL gradually merged with the Internet and dropped its subscription-only access, but its controlled access had followers, as well as detractors, who derided AOL as "the Internet with training wheels." Yahoo! (the company name included the exclamation point) was started in 1994 by two Stanford University students, Jerry Yang and David Filo, who initially compiled a guide to the Web manually. They took a crucial step that year by funding the site with advertisements, harking back to the Prodigy service. As the Web grew, it became less and less practical to continue indexing the Web manually, although automated indexing left much to be desired, because it had little intelligence to know whether the presence of a keyword implied that the Web site containing it actually contained useful information pertaining to that word.

The sites of Yahoo! and its competitors evolved to be more of a guide to the Web than just a search function, becoming what the press called a portal to the Web. As AOL and Prodigy had done in the pre-Web days, a visitor to a portal could find news, e-mail, discussion groups, winning lottery numbers, sports scores, and endless other information all in one place. The ease with which one could set up such a portal, drawing on content that others created, led to competitors, many of which were given absurdly high valuations by Wall Street in the late 1990s. After 2000, Yahoo! survived, while most of the others disappeared. AOL merged with the media giant Time-Warner in 2000 (actually it bought Time-Warner using its inflated stock valuation), but the merger did not work out. AOL has survived, though as a much smaller entity.

The emphasis on creating these portals did not fully solve the problem of searching for information. The portals included a search bar, but search was not among their priorities. One reason was that if a user found a site using a portal's search function, he or she would likely leave the portal and go to the sites found by the search, which meant this person was less likely to see the advertisements.[29] Into that breach stepped a few pure search engines. One of the most interesting of these was AltaVista, formed in 1995 as a subsidiary of the Digital Equipment Corporation, in part to take advantage of the high processing speeds of its own "Alpha" microprocessor. For a few brief years, AltaVista was the search engine that the most savvy Internet surfers preferred. It had a clean user interface, an ability to deliver results quickly, and a focus on search rather than other things. Digital Equipment , always a hardware company, failed to see that AltaVista was an extremely valuable property in its own right, not just as a way to sell Alpha chips, and as a result AltaVista faded as competitors entered the field.[30]

Google, like Yahoo! was founded by two Stanford University students, Larry Page and Sergei Brin. Stanford's location in the heart of Silicon Valley has given it an edge over MIT and other universities as places where ideas like these emerge, although there are many other factors as well, especially the relations Stanford has developed with venture capitalists located nearby. One factor seldom mentioned in the histories of Google is the presence on the Stanford faculty of Donald Knuth, whose book *The Art of Computer Programming, Volume Three: Sorting and Searching,* remains a classic work on the topic, decades after its publication in 1973.[31] The task set before search engines is a difficult one and can never be entirely automated. Histories of Google do mention the mentorship of Terry Winograd, a professor who had done pioneering work in artificial intelligence at the beginning of his career, when he developed a program called SHRDLU that exhibited an understanding of natural-language commands.[32] This is worth mentioning because it illustrates how far computing has evolved from its roots in World War II. Computers continue to serve as calculators, with a myriad of applications in science and engineering. Modern high-end supercomputers are the descendants of the ENIAC. Winograd and his colleagues working in artificial intelligence took early steps toward using a computer to process texts. Those steps faltered after initial modest success, but with Google, the efforts bore fruit.

Google surpassed the rival search engines because it consistently delivered better results when a user typed in a query. It ranked results not just by the number of times a term appeared on a Web page, for example, but also by how many other sites linked to a Web site. That was a Web version of a technique developed for scientific publishing by Eugene Garfield, in which the importance of a scientific paper was judged by how many other papers cited it in footnotes. Google's algorithm was more complex than that, and major portions of it remain a trade secret. Another factor contributing to Google's success was one we have encountered again and again: Google presented the user with a clean, simple search screen, devoid of pop-up ads, fancy graphics, and other clutter. Two other Web sites that consistently rank among the most visited, Wikipedia and Craigslist, also have a text-oriented design with few frills. That was a lesson that many Web sites have not learned; their creators apparently cannot resist the temptation to load a page with an angry fruit salad of too many colors, fonts, typefaces, and pop-up ads. These designs may seem a long way from the human factors work done during World War II on antiaircraft fire control devices, but they are the spiritual descendants.

Google surpassed the rival search engines because it consistently delivered better results when a user typed in a query. It ranked results not just by the number of times a term appeared on a Web page, for example, but also by how many other sites linked to a Web site. That was a Web version of a technique developed for scientific publishing by Eugene Garfield, in which the importance of a scientific paper was judged by how many other papers cited it in footnotes.

Facebook and Twitter

As of this writing, many press accounts of computing all feel obligated to mention Facebook and Twitter in the same sentence. But the two have little in common other than being currently very popular. Twitter, founded in 2006, is easy to explain. A messaging program that is restricted to a few lines of text, it is the product of the smart phone revolution and a creative response to the phones' small screens and keyboards. One can use Twitter on a laptop or desktop computer, but by using a full-sized screen and keyboard, blogs and other tools are more effective. A good percentage of the press coverage for Twitter is from journalists who are concerned that the program is going to adversely affect their careers—which may turn out to be true.

Facebook is another story, one with deep roots. Its meteoric rise was the subject of a Hollywood film, *The Social Network* (2010). One of the most interesting aspects of Facebook's success was how quickly it eclipsed the rival program MySpace, which as recently as 2008 was considered the established leader in such programs.[33] Facebook's overpowering presence suggests that the traditional view of the Internet and Web (in this context "traditional" means the way it was in 2005) is giving way. Many users log on to Facebook in the morning and leave it in a window (or on a separate screen) all day. It offers a single site for sharing gossip, news, photos, and many of the other offerings found on the Web. It is to the Web as Walmart is to retailing: a single place where someone can do all his or her shopping, under one roof, to the detriment of traditional specialty stores, and even supermarkets and department stores that had already integrated a lot of shopping. What is more, Facebook is beyond the reach of Google's search tentacles. One cannot locate, say, the enormous number of photographs stored on Facebook without joining and becoming an active member.

Before the Web transformed the Internet, AOL provided a one-stop access to similar services, using its private network of servers. After 1995 AOL was compelled to open up its private enclave, and it lost its privileged position as the favored portal to networked information. And before AOL there was Geocities, a Web site that used a geographical metaphor to organize chat rooms, personal Web sites, and other topics. And before Geocities there were bulletin boards and Usenet, with its cadre of Unix-savvy users trading information. Although seldom stated explicitly, a driving force for many of these online communities was sex, as noted by the obsession Facebook users have with the updates to one's "status." Historically this social force ranged from the simple expedient use of the network to find partners (of either sex) to date, to socialize with, to marry, or for seedier uses, including prostitution and pornography.[34] In 1990, as Usenet was establishing categories for discussion about the Unix operating system, an "alt" (for alternate) category was established, for topics not generally pertaining to computer programming or the network. Before long, "alt.sex" and "alt.drugs" appeared, prompting one user to state, "It was therefore artistically necessary to create alt.rock-n-roll, which I have also done."[35] The social forces driving AOL and the bulletin boards were the ancestors of the forces driving Facebook, Twitter, and similar programs in the twenty-first century. As with the invention of the personal computer itself, these forces drove networking from the bottom up, while privileged military and academic agencies drove networking from the top down. Today's world of networked computing represents a collision of the two.

7

CONCLUSION

Given all the press coverage and publicity, it seems appropriate to end this narrative with the rise of Facebook and Twitter as long as we do not lose sight of the general themes that have driven computing and computer technology from its origins. Above all, it is the force that drove the invention of the electronic computer in the first place: numerical calculation. Traditional number crunching, such as solving differential equations as the ENIAC did, remains a primary use of large computers in the scientific world. The early computers presented their results in numerical form, as tables of numbers; scientists increasingly can see the results in graphical form as well. Complementing the number-processing capabilities of modern mainframes is an ability to handle very large sets of data, using ever increasing capacities of disk storage and database software. This is an important point: behind many apps on a smart phone is a complex network of servers, routers, databases, fiber-optic cables, satellites, mass storage arrays, and sophisticated software.

Likewise, the current fascination with portable devices and Facebook should not obscure the continuing use of computers for business, the market that IBM exploited so well with its System/360 line in the late 1960s. Facebook may be the obsession right now, but corporations still use computers for prosaic applications like payroll and inventory, in many cases still programming in COBOL. Companies like FedEx employ networks of large mainframes to schedule their work and track shipments on the Internet. Customers can see where their package is by looking at their smart phone, behind which the company operates a network whose complexity is orders of magnitude greater than what IBM and its competitors offered a few decades ago. The federal government, including the Department of Defense, remains a huge user of computers of all sizes.

ARPA (now called DARPA) continues to drive innovation, although it may be hard-pressed to come up with something like the Internet again. With DARPA's support, the Defense Department has a counterpart to the Apple iPhone: small microprocessor-based devices that are embedded into weapons systems, such as unmanned aerial vehicles (UAVs), that use satellite data, on-board inertial guidance, and robotic vision to fly autonomously and hit remote targets. These are seamlessly linked to human commanders in the field. Some of these weapons are adapted from devices developed for consumer use, a reversal of the initial one-way flow of innovation from DARPA to the civilian world. The issue of intelligent, autonomous robotic weapons raises philosophical and moral questions about warfare that once were found only among science-fiction writers. This future is here, now.

We conclude by returning briefly to the four themes I identified at the outset, which I argue give a structure to the topic of computers and computing.

The digital paradigm has proved itself again and again as traditional methods of information handling and processing give way to manipulating strings of bits. It began with the pocket calculator replacing the slide rule. A few years ago, we saw the demise of film photography and the rise of digital cameras. Vinyl, analog recording has made a surprising comeback among music fans, but the current trend of downloading individual songs as digital files is not going to vanish. Airplanes are designed on computer screens and are flown by digital "fly-by-wire" electronics. After a few false starts, the e-book has established itself. A book is now a string of bits, delivered over a wireless network, to an e-book reader. Traditional books will not disappear, but will they occupy a niche like vinyl records? The list will go on.

The convergence of communications, calculation, data storage, and control also goes on, although with a twist. The handheld device can do all these things, but its communication function—voice telephone calls, texting, or Internet access—is split among different networks. Some of this is a product of the political and regulatory structure of the United States, and one finds a different mix in other countries. Future developments in this arena may depend as much on the U.S. Congress or the Federal Communications Commission as what comes from Apple's galactic headquarters in Cupertino, California.

Moore's law, the shorthand name for the steady advance of microelectronics, continues to drive computing. A recent study by the National Research Council hinted that although memory capacity continues to increase exponentially, the corresponding speed of microprocessor is not keeping pace.[1] The question of whether Moore's law illustrates the concept of technological determinism, so anathema to historians, remains a contested topic. The development of the personal computer refutes the deterministic thesis. So does the sudden rise of Facebook, even if a program like that requires a large infrastructure of silicon to be practical.

The fourth theme, the user interface, also remains central even as it evolves. No one disputes the importance of a clean, uncluttered design, and this has been noted regarding the success of Apple Computer's products and the most-visited Web sites, like Google. How do these issues relate to the design of antiaircraft weapons during World War II? That was when the discipline of operations research emerged. Operations research was a response to the realization that an engineer's job was not finished once a machine was designed; he or she then had to fit that device into a human context—in that case, of newly recruited soldiers and sailors who had little background in advanced technology yet who were being asked to operate sophisticated radar and other electronic devices.[2] The context has changed, but the underlying issue remains: computers are effective when their operation—and programming—is pushed down to the user level, whoever that user may be. And the plasticity of computers, the very quality that sets them apart from other machines, will always mean that their use will never be intuitive or obvious. As with the other themes, this too is evolving rapidly. It is startling to be reminded, for example, that many people today are accessing computers without using a mouse, never mind a QWERTY keyboard.

Zeno's paradox states that we can never truly understand computers and computing, at least until innovation in the field slows or stops. Perhaps if Moore's law slows down, software writers will have time to catch up with cleaner and less buggy programming. The only certainty is that the next few decades of the twenty-first century will be as dynamic and effervescent as the decades since 1945.

NOTES

Introduction

1. Paul Ceruzzi, *A History of Modern Computing, 2nd ed.* (Cambridge, MA: MIT Press, 2003.

Chapter 1

1. Bernard O. Williams, "Computing with Electricity, 1935–1945" (Ph.D. diss., University of Kansas, 1984, University Microfilms 8513783), 310.
2. Tom Standage, *The Victorian Internet: The Remarkable Story of the Telegraph and the Nineteenth Century's On-Line Pioneers* (New York: Walker, 1998).
3. Brian Randell, ed., *The Origins of Digital Computers: Selected Papers 2nd ed.* (Berlin: Springer-Verlag 1975); Charles and Ray Eames, *A Computer Perspective: Background to the Computer Age, New Edition* (Cambridge, MA: Harvard University Press, 1990); Herman H. Goldstine, *The Computer from Pascal to von Neumann* (Princeton, NJ: Princeton University Press, 1972). Also note the founding of the Charles Babbage Institute Center for the History of Information Technology, in 1978.
4. Dag Spicer, "Computer History Museum Report," *IEEE Annals of the History of Computing* 30, no. 3 (2008): 76–77.
5. U.S. National Security Agency, "The Start of the Digital Revolution: SIGSALY: Secure Digital Voice Communications in World War II" (n.d.), available from the National Cryptologic Museum, Fort Meade, MD.
6. Edison was one of several inventors who developed similar devices around 1870. The initial Morse telegraph printed the dots and dashes of the code, but that apparatus was abandoned after operators found that with training, they could transcribe the sounds more rapidly.
7. "Edward Kleinschmidt: Teletype Inventor," *Datamation* (September 1977): 272–273.
8. Jo Ann Yates, *Control through Communication* (Baltimore, MD: Johns Hopkins University Press, 1989); see also James R. Beniger, *The Control Revolution: Technological and Economic Origins of the Information Society* (Cambridge, MA: Harvard University Press, 1986) .
9. Arthur L. Norberg, "High Technology Calculation in the Early Twentieth Century: Punched Card Machinery in Business and Government," *Technology and Culture* 31 (October 1990): 753–779; also James W. Cortada, *Before the*

Computer: IBM, NCR, Burroughs, and Remington Rand and the Industry They Created, 1865–1956 (Princeton, NJ: Princeton University Press, 1993.
10. Lars Heide, *Punched-Card Systems in the Early Information Explosion, 1880–1945* (Baltimore, MD: Johns Hopkins University Press, 2009).
11. Thomas P. Hughes, *Networks of Power: Electrification in Western Society, 1880–1930* (Baltimore, MD: Johns Hopkins University Press, 1983).

Chapter 2

1. Konrad Zuse, *The Computer—My Life*, English translation of *Der Computer, Mein Lebenswerk* (New York: Springer-Verlag, 1993), 44. Originally published in German in 1962.
2. Ibid., p. 46.
3. Alan M. Turing, "On Computable Numbers, with an Application to the Entscheidungsproblem," *Proceedings of the London Mathematical Society*, 2nd series, 42 (1936): 230–265.
4. Constance Reid, *Hilbert* (New York: Springer-Verlag, 1970), 74–83; Charles Petzold, *The Annotated Turing* (Indianapolis, IN: Wiley, 2008), 35–53. Petzold traces (p. 31)Turing's paper back to Hilbert's "tenth problem," the tenth in a list of twenty-three that in 1900 Hilbert suggested would challenge mathematicians of the coming century.
5. Petzold, *Annotated Turing*.
6. Charles Babbage, *Passages from the Life of a Philosopher* (London, 1864), reprinted in Charles Babbage Institute, *Babbage's Calculating Engines* (Los Angeles: Tomash Publishers, 1982), 170–171.
7. Zuse's simultaneous discovery was little known until the publication of his memoir, *Der Computer, Mein Lebenswerk*, in 1962 in German. His pioneering hardware work was somewhat better known beginning in the 1950s, when he began manufacturing and selling commercial computers, mainly in West Germany.
8. William Aspray, *John von Neumann and the Origins of the Modern Computing* (Cambridge, MA: MIT Press, 1990), 178–180.
9. It is difficult to avoid using anthropomorphic terms like *memory* or *read*, and these can be misleading. I use them sparingly in this chronicle.
10. Wallace J. Eckert, *Punched Card Methods in Scientific Computation* (New York: Thomas J. Watson Astronomical Computing Bureau, 1940).
11. Paul Ceruzzi, "Crossing the Divide: Architectural Issues and the Emergence of the Stored Program Computer, 1935–1955," *IEEE Annals of the History of Computing* 19, no. 1 (1997): 5–12.

12. Howard Aiken, "Proposed Automatic Calculating Machine," written in 1937, and published in *IEEE Spectrum* (August 1964): 62–69.
13. Harvard University Computation Laboratory, *A Manual of Operation for the Automatic Sequence Controlled Calculator* (Cambridge, MA: Harvard University 1946).
14. Alice R. Burks and Arthur W. Burks, *The First Electronic Computer: The Atanasoff Story* (Ann Arbor: University of Michigan Press, 1988).
15. Brian Randell, "The Colossus," in N. Metropolis, J. Howlett, and Gian-Carlo Rota, eds., *A History of Computing in the Twentieth Century* (New York: Academic Press, 1980), 47–92.
16. Samuel S. Snyder, "Computer Advances Pioneered by Cryptologic Organizations," *Annals of the History of Computing* 2 (1980): 60–70.
17. Perry Crawford Jr., "Instrumental Analysis in Matrix Algebra" (bachelor's thesis, MIT, 1939); Perry Crawford Jr., "Automatic Control by Arithmetic Operations" (master's thesis, MIT, 1942).
18. Claude E. Shannon, "A Symbolic Analysis of Relay and Switching Circuits," *Transactions of the American Institution of Electrical Engineers* 57 (1938): 713–723.
19. Norbert Wiener, *Cybernetics, or Control and Communication in the Animal and the Machine* (Cambridge, MA: MIT Press, 1948).
20. Vannevar Bush, "As We May Think," *Atlantic Monthly* 176 (1945): 101–108.
21. Burks and Burks, *The First Electronic Computer*.

Chapter 3

1. John von Neumann, "First Draft of a Report on the EDVAC," (Philadelphia, PA: Moore School of Electrical Engineering, University of Pennsylvania, June 30, 1945).
2. Kent C. Redmond and Thomas M. Smith, *Project Whirlwind: the History of a Pioneer Computer* (Bedford, MA: Digital Press, 1980); also see Atsushi Akera, *Calculating a Natural World: Scientists, Engineers, and Computers during the Rise of U.S. Cold War Research* (Cambridge, MA: MIT Press, 2007).
3. Alan Borning, "Computer Reliability and Nuclear War," *Communications of the ACM* 30/2 (1987): 112–131.
4. Maurice V. Wilkes, Memoirs of a Computer Pioneer Cambridge, MA: MIT Press, 1985; also Grace Murray Hopper, "Compiling Routines," *Computers and Automation* 2 (May 1953): 1–5.
5. Ernest Braun and Stuart McDonald, *Revolution in Miniature: The History and Impact of Semiconductor Electronics*, 2nd ed. (Cambridge: Cambridge University Press, 1982).

6. Ibid.
7. Jamie Parker Pearson, ed., *Digital at Work* (Bedford, MA: Digital Press, 1992), 10–11.
8. Ibid., 21.
9. Joseph November, *Digitizing Life: The Rise of Biomedical Computing in the United States* (Baltimore, MD: Johns Hopkins University Press, 2011).
10. Chigusa Kita, "J.C.R. Licklider's Vision for the IPTO," *IEEE Annals of the History of Computing* 25, no. 3 (2003): 62–77.

Chapter 4

1. Mara Mills, "Hearing Aids and the History of Electronics Miniaturization," *IEEE Annals of the History of Computing* 33, no. 2 (2011): 24–44; James Phinney Baxter III, *Scientists against Time* (New York: Little, Brown and Co., 1946).
2. Leslie Berlin, *The Man behind the Microchip: Robert Noyce and the Invention of Silicon Valley* (New York: Oxford University Press, 2005).
3. T. A. Wise, "IBM's $5,000,000,000 Gamble," *Fortune* (September 1966), pp. 118–123, 224, 226, 228.

Chapter 5

1. The UNIVAC had about 5,000 vacuum tubes and several thousand solid-state diodes, capacitors, and resistors. It stored data as acoustic pulses in tubes of mercury.
2. Gordon E. Moore, "Cramming More Components onto Integrated Circuits," *Electronics*, April 19, 1965, 114–117. The cartoon, which appears on page 116, is attributed to Grant Compton.
3. David Hounshell, *From the American System to Mass Production, 1800–1932: The Development of Manufacturing Technology in the United States* (Baltimore, MD: Johns Hopkins University Press, 1984). Hounshell points out that the Model T in fact did undergo a number of changes during its production run, but he does acknowledge the existence of a "Model T dilemma."
4. Adi J. Khambata, *Introduction to Large-Scale Integration* (New York: Wiley-Interscience, 1969), 81–82.
5. John L. Hennessy and David A. Patterson, *Computer Architecture: A Quantitative Approach* (San Mateo, CA: Morgan Kaufmann, 1990).
6. M .V. Wilkes, "The Best Way to Design an Automatic Calculating Machine," in *Computer Design Development: Principal Papers*, ed. Earl E. Swartzlander Jr., 266–270 (Rochelle Park, NJ: Hayden, 1976).
7. Gordon E. Moore, "Microprocessors and Integrated Electronic Technology," *Proceedings of the IEEE* 64 (1976): 837–841.

8. Charles J. Bashe, Lyle R. Johnson, John H. Hunter, and Emerson W. Pugh, *IBM's Early Computers* (Cambridge, MA: MIT Press 1986), 505–513. IBM's internal code name for the 1620 was "CADET," which some interpreted to mean "Can't Add; Doesn't Even Try"!

9. Paul Freiberger and Michael Swaine, *Fire in the Valley: The Making of the Personal Computer* (Berkeley, CA: Osborne/McGraw-Hill, 1984); also Ross Knox Bassett, *To the Digital Age: Research Labs, Start-Up Companies, and the Rise of MOS Technology* (Baltimore, MD: Johns Hopkins University Press, 2002).

10. A KIM-1 was my first computer.

11. Bruce Newman, "Apple's Third Founder Refuses to Submit to Regrets," *Los Angeles Times*, June 9, 2010.

12. Fred Turner, *From Counterculture to Cyberculture: Stewart Brand, the Whole Earth Network, and the Rise of Digital Utopianism* (Chicago: University of Chicago Press, 2006).

13. Vannevar Bush, "As We May Think," *Atlantic Monthly* 176 (July 1945): 101–108.

14. Robert M. Metcalfe, "How Ethernet Was Invented," *IEEE Annals of the History of Computing* 16 (1994): 81–88.

Chapter 6

1. Lawrence G. Roberts, "The ARPANET and Computer Networks," in *A History of Personal Workstations*, ed. Adele Goldstine, 143–171 (New York: ACM Press, 1988); Daniel P. Siewiorek, C. Gordon Bell, and Allen Newell, *Computer Structures: Principles and Examples* (New York: McGraw-Hill, 1982), 396–397, and their *Computer Structures: Readings and Examples* (New York: McGraw-Hill, 1971), 510–512.

2. Janet Abbate, *Inventing the Internet* (Cambridge, MA: MIT Press, 1999).

3. William Aspray and Paul Ceruzzi, eds., *The Internet and American Business* (Cambridge, MA: MIT Press, 2008).

4. U.S. Department of Commerce, National Telecommunications and Information Administration, "Domain Names: U.S. Principles on the Internet's Domain Name and Addressing System," http://www.ntia.doc.gov/other-publication/2005/us-principles-internets-domain-name-and-addressing-system. (accessed December 29, 2011).

5. Alfred Glossbrenner, *The Complete Handbook of Personal Computer Communications*, 3rd ed. (New York: St. Martin's Press, 1990), xiv.

6. Stewart Brand, ed., *Whole Earth Software Catalog* (New York: Quantum Press/Doubleday, 1984), 139.

7. Ibid., 140.

8. Ibid., 144.
9. CompuServe databases were handled by Digital Equipment Corporation PDP-10s and its successor computers, which used octal arithmetic to address data. CompuServe users' account numbers thus contained the digits 0 through 7 but never 8 or 9.
10. Glossbrenner, *Complete Handbook*, 68 (emphasis added).
11. Kara Swisher, *AOL.COM: How Steve Case Beat Bill Gates, Nailed the Netheads, and made Millions in the War for the Web* (New York: Times Business Random House, 1998). chap. 2.
12. Ibid.
13. Ibid.; Alfred Glossbrenner, *The Complete Handbook of Personal Computer Communications* (New York: St. Martin's Press, 1983).
14. Siewiorek et al., *Computer Structures*, 387–438; Brand, *Whole Earth Software Catalog*, 138–157.
15. Ibid. Quarterman states that the term is an acronym for "Because It's Time NETwork"; other accounts say it stood for "Because It's There NETwork," referring to the existence of these large IBM installations already in place.
16. The Smithsonian Institution, for example, had BITNET accounts for its employees long before most of these employees had an Internet connection.
17. The initial NSF backbone used networking software called "fuzzball," which was not carried over to the T1 network.
18. "NSFNET—National Science Foundation Network," Living Internet, online resource, at <www.livinginternet.com/i/ii_nsfnet.htm> (accessed November 10, 2005); also Jay P. Kesan and Rajiv C. Shah, "Fool Us Once, Shame on You—Fool Us Twice, Shame on Us: What We Can Learn from the Privatization of the Internet Backbone Network and the Domain Name System," *Washington University Law Quarterly* 79 (2001), p. 106.
19. Ed Krol, *The Whole Internet User's Guide and Catalog* (Sebastopol, CA: O'Reilly & Associates,1992), appendix C.
20. Robert Kahn, personal communication with the author.
21. This information was taken from the official web page of Congressman Rick Boucher www.boucher.house.gov, accessed June 2006. Boucher has since left the House of Representatives, and the web page is no longer active. It may however be found through the Internet Archive (www.archive.org).
22. 42 U.S.C. 1862, paragraph g.
23. Stimson Garfinkel, "Where Streams Converge," *Hot Wired*, September 11, 1996.
24. Tim Berners-Lee, *Weaving the Web: The Original Design and Ultimate Destiny of the World Wide Web* (San Francisco: HarperCollins, 1999).

25. Ibid.
26. Shane Greenstein, "Innovation and the Evolution of Market Structure for Internet Access in the United States," in *The Internet and American Business*, ed. William Aspray and Paul Ceruzzi, 47–103. Cambridge, MA: MIT Press, 2008.
27. "Martin Cooper: Inventor of Cell Phones says they're now 'Too Complicated,'" *Huffington Post* , March 18, 2010. http://www.huffingtonpost.com/2009/11/06/martin-cooper-inventor-of_n_348146.html.
28. Joshua Cuneo, "'Hello Computer': The Interplay of Star Trek and Modern Computing," in *Science Fiction and Computing: Essays on Interlinked Domains*, ed. David L. Ferro and Eric G. Swedin, 131–147 (Jefferson, NC: McFarland, 2011).
29. Thomas Haigh, "The Web's Missing Links: Search Engines and Portals," in The *Internet and American Business*, ed. William Aspray and Paul Ceruzzi (Cambridge, MA: MIT Press 2008), chap. 5.
30. C. Gordon Bell, "What Happened? A Postscript," in Edgar H. Schein, *DEC is Dead: Long Live DEC: The lasting Legacy of Digital Equipment Corporation* (San Francisco: Berrett-Koehler Publishers, 2003): 292–301.
31. Donald E. Knuth, *The Art of Computer Programming, Vol. 3: Sorting and Searching* (Reading, MA: Addison-Wesley, 1973).
32. Terry Winograd, *Understanding Natural Language* (New York: Academic Press, 1972).
33. Christine Ogan and Randall A. Beam, "Internet Challenges for Media Businesses," in *The Internet and American Business*, ed. William Aspray and Paul E. Ceruzzi, chap. 9. (Cambridge, MA: MIT Press 2008).
34. Blaise Cronin, "Eros Unbound: Pornography and the Internet," in *The Internet and American Business*, ed. William Aspray and Paul E. Ceruzzi, chap. 15 (Cambridge, MA: MIT Press, 2008).
35. Brian Reid, quoted in Peter H. Salus, *Casting the Net: From ARPANET to INTERNET and Beyond* (Reading, MA: Addison-Wesley, 1995), 147.

Chapter 7

1. Samuel H. Fuller and Lynette I. Millett, *The Future of Computing: Game Over or Next Level?* (Washington, DC: National Academies Press, 2011).
2. Agatha C. Hughes and Thomas P. Hughes, eds., *Systems, Experts, and Computers: The Systems Approach in Management and Engineering, World War II and After* (Cambridge, MA: MIT Press, 2000) chaps. 1 and 2.

Glossary

1. Donald Knuth, *The Art of Computer Programming*, Vol. 1: Fundamental Algorithms (Reading, MA: Addison-Wesley, 1969, 4).

GLOSSARY

Algorithm

According to Donald Knuth, "A finite set of rules, which gives a sequence of operations for solving a specific type of problem."[1] Like a recipe to cook a dish, only every step is precisely defined so that a machine can carry the steps out. Implied is that these steps be completed in a finite amount of time.

ARPA

Advanced Research Projects Agency, established by the U.S. Defense Department in 1968. A research agency not affiliated with the specific services, which has been charged with advanced research not necessarily tied to a specific weapons system. The initial acronym was later changed to DARPA.

Artificial Intelligence (AI)

The classic definition is that of a computer that performs actions that, if done by a human being, would be considered intelligent. In practice, as computers become more powerful, actions such as playing a good game of chess are no longer considered AI, although they once were considered at the forefront of AI research. At the same time, despite decades of research and advances in technology, no computer has an ability to converse with a human being in a normal fashion on a variety of topics.

Binary

A representation of numbers or values where only two values are allowed. This can be, for example, 1 or 0 in arithmetic, yes or no in logic, on or off for an electrical switch, or the presence of absence of a current in a wire. Although less comfortable for humans who are accustomed to decimal (base 10) numbers, binary has overwhelming advantages from both an engineering and a theoretical standpoint.

Bit

A binary digit, either 1 or 0.

Boolean algebra
Like binary arithmetic, a system of logic that accepts only two values: true or false. The rules for manipulating these values are congruent to the rules for performing arithmetic in the binary system, with the numbers 1 and 0, with minor differences.

Byte
Eight bits, treated as a single unit. A sequence of eight bits is enough to encode the letters of the Roman alphabet in upper and lower case; the numerals, punctuation, and other symbols; and other special characters that may be used, for example, to control a printer. The typical measure of data storage capacity; for example, 100 Megabytes would be equal to 800 million bits of data.

Calculator
A mechanical or electronic device that evaluates the four functions of arithmetic: addition, subtraction, multiplication, and division.

Chatroom (or Chat room)
A virtual space found on computer networks in which a person types a message, which is read by all others who are logged into that space. They may in turn reply. The earliest chatrooms allowed no more than simple text. That was later augmented by voice and simple graphics, but basic text dominated. Chatrooms continue to be used, although social media services like Facebook have largely supplanted them.

Client-server
An arrangement of networked computers, in which "clients" (workstations or personal computers) receive data from large-capacity computers called "servers." The data are typically in raw form, while the servers use their computing capacity to handle housekeeping, graphics, and other computations.

Compiler
A specialized computer program that accepts as input commands written in a form familiar to humans and produces as output instructions that a machine can execute, usually in the form of strings of binary numbers.

Computer
The definition of this term has changed over the years, but it generally refers to a machine, almost always using electronic components, that performs calculations, stores data, and carries out sequences of operations automatically. The modern definition assumes that the program that directs the computer's operation is also stored internally in its memory, along with the data.

Control
In this context, the part of a computer that directs the other circuits to perform calculation, storage, input, and output as it decodes the instructions of a program.

Data
From the Latin word (plural; singular *datum*) for "things that are given." Any information, in coded form, that a computer can process. Although derived from a plural word, *data* is often treated as a singular noun in English.

Electromechanical
A method of switching or manipulating electrical currents, in which the switching is done by metal contacts, which in turn are activated by electrical currents. These include so-called relays, which were once common in telephone switching, and the stepping switching of decimal wheels in a punched-card machine. Contrast with *electronic*, in which all the switching is done by electrons in either a vacuum or solid-state device.

Electronic
A method of switching that uses electrons moving at high speeds. No mechanical devices are used. Early electronic computers used vacuum tubes, later solid-state transistors and integrated circuits.

Graphical User Interface (GUI)
A method of interacting with a computer by clicking a mouse on symbolic information presented on a video screen. For most consumers, this has replaced the earlier method of directly typing in a command, such as "print" or "save."

Integrated circuit
An electronic device in which all of the classic components of a circuit, such as resistors, transistors, capacitors, and their connecting wires, are combined on a single piece of material, usually silicon. Often called a *microchip*, or simply *chip*.

Internet
A term initially meant to designate a network of heterogeneous networks. The current meaning is of the worldwide network that uses the TCP/IP protocols, and conforms to the addressing of the Domain Name System, as administered by a governing body known as the Internet Corporation for Assigned Names and Numbers (ICANN).

LISP
List Processing, a computer language once favored by artificial intelligence researchers.

Memory
The part of a computer where data is stored. Computers have a hierarchy of memory devices, with a small memory, of lesser capacity, that stores and retrieves data at high speed, followed by slower but higher-capacity devices, such as spinning disks or magnetic tape drives. The anthropomorphic implications can be misleading, as computer memories operate on fundamentally different principles from human memory.

Microcode
Detailed computer programs that direct the detailed operations of a processor. Typically stored on read-only memory. *See* read-only memory.

Microprocessor
A device that contains most of the basic components of a general-purpose stored program computer on a single chip; typically used in conjunction with random-access memory and read-only memory chips. *See* random-access memory; read-only memory.

MOS
Metal-oxide semiconductor: a type of integrated circuit that lends itself to high densities and low power consumption.

Mouse
A device that allows a computer user to select items on a screen.

Operating System
A specialized program that manages the housekeeping chores of a computer, such as transferring data from its internal memory to a disk or to a terminal and interpreting clicks of a mouse or keystrokes.

Packet switching
A method of transmitting data over a computer network by breaking up a file into smaller sections, called packets, each packet also containing a header that gives information about the packet's destination and contents. The technical basis for the Internet.

Program
A sequence of instructions executed by a computer to perform actions desired by its user. See also *software*.

Protocols
The conventions that govern the switching of packets. Similar to the addressing of a letter, which has conventions as to the placement of the destination and return address, the placement and amount of postage, and so on, regardless of the contents of the letter.

Random-access memory (RAM)
That portion of a computer's memory that has the highest access speeds, and often at less capacity than a slower disk. The term is a misnomer, since it originally referred to a disk, in which the time it took to access a datum depended on the random placement of that information on the disk. It now refers to the internal memory, the access of which is nearly the same regardless of where it is placed.

Read-only memory (ROM)
That portion of a computer's memory that stores data that can be read, but not changed by the user. Typically a read-only memory chip stores programming (*see* microcode) that tailors the microprocessor to execute specific functions.

Relay
A mechanical switch that is activated by an electromagnet. The term came from telegraphy, in which a telegraph signal is sent over long distances by being freshly regenerated at relay stations, as a baton is passed by relay racers during a long race.

Server
A computer that typically contains a large memory and that can deliver data over networks at high speeds. A modern descendant of the earlier model of mainframe computers that were time-shared.

Silicon
An element, atomic number 14 on the periodic table, which has properties that are well suited for the construction of integrated circuits.

Software
The suite of programs, including applications, operating systems, and system programs, that a computer executes.

Symbolic logic
See Boolean algebra.

TCP/IP
Transmission Control Protocol/Internet Protocol. *See* protocols.

Time-sharing
A method of using a large computer connected to a number of terminals. Because of the speeds of the computer in relation to the reaction time of the user, a person seated at a terminal has the impression that her or she has direct, sole access to the computer. (*See also* client-server.) An arrangement of networked computers that combines machines with high memory capacity and switching speeds with intelligent terminals.

Transistor
An electronic device that switches in a solid piece of material, typically silicon or germanium.

Vacuum tube
A device that switches electrons that move in a vacuum, excited by a hot filament.

Word
The set of binary digits processed together by a computer. A typical modern personal computer has a word length of 32 or 64 bits.

Workstation
A high-end personal computer with rich graphics, networking, and calculating ability.

World Wide Web
A program available on the Internet that allows those connected to the network to access information easily, whether it is stored locally or on another continent and regardless of the particular computer or server where it is located.

FURTHER READING

Abbate, Janet. *Inventing the Internet*. Cambridge, MA: MIT Press, 1999.

Aspray, William, and Paul E. Ceruzzi, eds. *The Internet and American Business*. Cambridge, MA: MIT Press, 2008.

Berlin, Leslie. *The Man behind the Microchip: Robert Noyce and the Invention of Silicon Valley*. New York: Oxford University Press, 2005.

Berners-Lee, Tim. *Weaving the Web: The Original Design and Ultimate Destiny of the World Wide Web by Its Inventor*. San Francisco: Harper, 1999.

Beyer, Kurt. *Grace Hopper and the Invention of the Information Age*. Cambridge, MA: MIT Press, 2009.

Braun, Ernest, and Stuart Macdonald. *Revolution in Miniature: The History and Impact of Semiconductor Electronics Re-explored in an Updated and Revised Second Edition*. Cambridge: Cambridge University Press, 1982.

Brooks, Frederick P., Jr. *The Mythical Man-Month: Essays on Software Engineering*. 2nd ed. Reading, MA: Addison-Wesley, 1995.

Campbell-Kelly, Martin. *From Airline Reservations to Sonic the Hedgehog: A History of the Software Industry*. Cambridge, MA: MIT Press, 2003.

Ceruzzi, Paul E. *A History of Modern Computing*. 2nd ed. Cambridge, MA: MIT Press, 2003.

Ensmenger, Nathan. *The Computer Boys Take Over: Computers, Programmers, and the Politics of Technical Experise*. Cambridge, MA: MIT Press, 2010.

Freiberger, Paul, and Michael Swaine. *Fire in the Valley: The Making of the Personal Computer*. Berkeley, CA: Osborne/McGraw-Hill, 1984.

Glossbrenner, Alfred. *The Complete Handbook of Personal Computer Communications: The Bible of the Online World*. 3rd ed. New York: St. Martin's Press, 1990.

Goldstein, Herman H. *The Computer from Pascal to von Neumann*. Princeton: Princeton University Press, 1972.

Grier, David Alan. *When Computers Were Human*. Princeton, NJ: Princeton University Press, 2005.

Hanson, Dirk. *The New Alchemists: Silicon Valley and the Microelectronics Revolution*. Boston: Little, Brown, 1982.

Kidder, Tracy. *The Soul of a New Machine*. Boston: Little, Brown, 1981.

Lécuyer, Christophe, and David Brock. *Makers of the Microchip: A Documentary History of Fairchild Semiconductor*. Cambridge, MA: MIT Press, 2010.

Levy, Steven. *Hackers: Heroes of the Computer Revolution*. New York: Anchor Doubleday, 1984.

Mahoney, Michael S. *Histories of Computing*. Cambridge, MA: Harvard University Press, 2011.

Manes, Stephen, and Paul Andrews. *Gates: How Microsoft's Mogul Reinvented an Industry, and Made Himself the Richest Man in America*. New York: Doubleday, 1993.

Murray, Charles. *The Supermen: The Story of Seymour Cray and the Technical Wizards behind the Supercomputer*. Hoboken, NJ: Wiley, 1997.

Nelson, Theodor. *Computer Lib*. South Bend, IN: Ted Nelson, 1974.

Norberg, Arthur L., and Judy O'Neill. *Transforming Computer Technology: Information Processing for the Pentagon, 1962–1986*. Baltimore, MD: Johns Hopkins University Press, 1996.

Pugh, Emerson. *Building IBM: Shaping an Industry and Its Technology*. Cambridge, MA: MIT Press, 1995.

Randell, Brian, ed. *The Origins of Digital Computers: Selected Papers*. 2nd ed. Berlin: Springer-Verlag, 1975.

Redmond, Kent C., and Thomas M. Smith. *Project Whirlwind: The History of a Pioneer Computer*. Bedford, MA: Digital Press, 1980.

Reid, T. R. *The Chip: How Two Americans Invented the Microchip and Launched a Revolution*. New York: Simon & Schuster, 1985.

Rojas, Raul, and Ulf Hashhagen, eds. *The First Computers: History and Architectures*. Cambridge, MA: MIT Press, 2000.

Salus, Peter. *A Quarter Century of UNIX*. Reading, MA: Addison-Wesley, 1994.

Smith, Douglas, and Robert Alexander. *Fumbling the Future: How Xerox Invented, Then Ignored, the First Personal Computer*. New York: Morrow, 1988.

Torvalds, Linus, and David Diamond. *Just for Fun: The Story of an Accidental Revolutionary*. New York: HarperCollins, 2001.

Waldrop, M. *Mitchell. The Dream Machine: J.C.R. Licklider and the Revolution That Made Computing Personal*. New York: Viking Press, 2001.

Watson, Thomas, Jr. *Father, Son & Co.: My Life at IBM and Beyond*. New York: Bantam Books, 1990.

Weizenbaum, Joseph. *Computer Power and Human Reason*. San Francisco: Freeman, 1976.

Wolfe, Tom. "The Tinkerings of Robert Noyce." *Esquire* (December 1983): 346–374.

Yates, JoAnne. *Control through Communication: The Rise of System in American Management*. Baltimore, MD: Johns Hopkins University Press, 1989.

Zachary, G. *Pascal. Endless Frontier: Vannevar Bush, Engineer of the American Century*. New York: Free Press, 1997.

Zuse, Konrad. *The Computer—My Life*. Berlin: Springer-Verlag, 1993.

BIBLIOGRAPHY

Abbate, Janet. *Inventing the Internet*. Cambridge, MA: MIT Press, 1999.

Akera, Atsushi. *Calculating a Natural World: Scientists, Engineers, and Computers during the Rise of U.S. Cold War Research*. Cambridge, MA: MIT Press, 2007.

Aspray, William. *John von Neumann and the Origins of Modern Computing*. Cambridge, MA: MIT Press, 1990.

Aspray, William, ed. *Computing before Computers*. Ames: Iowa State University Press, 1990.

Aspray, William, and Paul E. Ceruzzi, eds. *The Internet and American Business*. Cambridge, MA: MIT Press, 2008.

Bardini, Thierry. *Bootstrapping: Douglas Engelbart, Coevolution, and the Origins of Personal Computing*. Stanford, CA: Stanford University Press, 2000.

Bashe, Charles J., Lyle R. Johnson, John H. Palmer, and Emerson W. Pugh. *IBM's Early Computers*. Cambridge, MA: MIT Press, 1986.

Bassett, Ross Knox. *To the Digital Age: Research Labs, Start-up Companies, and the Rise of MOS Technology*. Baltimore, MD: Johns Hopkins University Press, 2002.

Baxter, James Phinney III. Scientists against Time. New York: Little, Brown and Co., 1946.

Bell, C. Gordon, and Allen Newell. *Computer Structures: Readings and Examples. McGraw-Hill Computer Science Series*. New York: McGraw-Hill, 1971.

Beniger, James R. *The Control Revolution: Technological and Economic Origins of the Information Society*. Cambridge, MA: Harvard University Press, 1986.

Berkeley, Edmund. *Giant Brains, or Machines That Think*. New York: Wiley, 1949.

Berlin, Leslie. *The Man behind the Microchip: Robert Noyce and the Invention of Silicon Valley*. New York: Oxford University Press, 2005.

Berners-Lee, Tim. *Weaving the Web: The Original Design and Ultimate Destiny of the World Wide Web by Its Inventor*. San Francisco: Harper, 1999.

Borning, Alan. "Computer Reliability and Nuclear War." *Communications of the ACM* 30 (2) (1987): 112–131.

Brand, Stewart. "Spacewar: Fanatic Life and Symbolic Death among the Computer Bums." *Rolling Stone*, December 7, 1972, 50–58.

Braun, Ernest, and Stuart Macdonald. *Revolution in Miniature: The History and Impact of Semiconductor Electronics Re-explored in an Updated and. Revised Second Edition*. Cambridge: Cambridge University Press, 1982.

Brooks, Frederick P., Jr. *The Mythical Man-Month: Essays on Software Engineering*. Reading, MA: Addison-Wesley, 1975.

Burks, Alice Rowe. *Who Invented the Computer? The Legal Battle That Changed Computing History*. Amherst, NY: Prometheus Books, 2003.

Burks, Alice R., and Arthur W. Burks. *The First Electronic Computer: The Atanasoff Story*. Ann Arbor: University of Michigan Press, 1988.

Bush, Vannevar. "As We May Think." *Atlantic Monthly*, 176 (July 1945): 101–108.

Campbell-Kelly, Martin. *From Airline Reservations to Sonic the Hedgehog: A History of the Software Industry*. Cambridge, MA: MIT Press, 2003.

Ceruzzi, Paul. "Crossing the Divide: Architectural Issues and the Emergence of the Stored Program Computer, 1935–1955." *IEEE Annals of the History of Computing* 19 (1) (1997): 5–12.

Ceruzzi, Paul E. *A History of Modern Computing*. 2nd ed. Cambridge, MA: MIT Press, 2003.

Charles Babbage Institute. *Babbage's Calculating Engines*. Los Angeles: Tomash Publishers, 1982.

Cohen, I. Bernard. *Howard Aiken: Portrait of a Computer Pioneer*. Cambridge, MA: MIT Press, 1999.

Cortada, James W. *Before the Computer: IBM, NCR, Burroughs, and Remington Rand and the Industry they Created, 1865–1956*. Princeton, NJ: Princeton University Press, 1993.

Crawford, Perry Jr. "Instrumental Analysis in Matrix Algebra." Bachelor's thesis, MIT, 1939.

Cronin, Blaise. "Eros Unbound: Pornography and the Internet." In *The Internet and American Business*, ed. William Aspray and Paul E. Ceruzzi, chap. 15. Cambridge, MA: MIT Press, 2008.

Cuneo, Joshua "'Hello Computer': The Interplay of Star Trek and Modern Computing." In *Science Fiction and Computing: Essays on Interlinked Domains*, ed. David L. Ferro and Eric G. Swedin, 131–147. Jefferson, NC: McFarland, 2011.

Eames, Charles, and Ray Eames. *A Computer Perspective: Background to the Computer Age*. Cambridge, MA: Harvard University Press, New Edition, 1990.

Eckert, Wallace. *Punched Card Methods in Scientific Computation*. New York: Thomas J. Watson Astronomical Computing Bureau, 1940.

"Edward Kleinschmidt: Teletype Inventor." *Datamation* (September 1977): 272–273.

Engineering Research Associates. *High Speed Computing Devices*. New York: McGraw-Hill, 1950.

Freiberger, Paul, and Michael Swaine. *Fire in the Valley: The Making of the Personal Computer*. Berkeley, CA: Osborne/McGraw-Hill, 1984.

Fuller, Samuel H., and Lynette I. Millett. *The Future of Computing: Game Over or Next Level?* Washington, DC: National Academies Press, 2011.

Garfinkel, Stimson. "Where Streams Converge." *Hot Wired*, September 11, 1996.

Glossbrenner, Alfred. *The Complete Handbook of Personal Computer Communications: The Bible of the Online World*. New York: St. Martin's Press, 1983.

Glossbrenner, Alfred. *The Complete Handbook of Personal Computer Communications: The Bible of the Online World*. 3rd ed. New York: St. Martin's Press, 1990.

Goldberg, Adele, ed. *A History of Personal Workstations*. Reading, MA: Addison-Wesley, 1988.

Goldstine, Herman H. *The Computer from Pascal to von Neumann*. Princeton, NJ: Princeton University Press, 1972.

Grier, David Alan. *When Computers Were Human*. Princeton, NJ: Princeton University Press, 2005.

Hanson, Dirk. *The New Alchemists: Silicon Valley and the Microelectronics Revolution*. Boston: Little, Brown, 1982.

Harvard University Computation Laboratory. *A Manual of Operation for the Automatic Sequence Controlled Calculator*. Cambridge, MA: Harvard University, 1946.

Heide, Lars. *Punched-Card Systems in the Early Information Explosion, 1880–1945*. Baltimore, MD: Johns Hopkins University Press, 2009.

Hennessy, John L. and David A. Patterson. *Computer Architecture: A Quantitative Approach*. San Mateo, CA: Morgan Kaufmann, 1990.

Hopper, Grace Murray. "Compiling Routines." *Computers and Automation* 2 (May 1953): 1–5.

Hounshell, David. *From the American System to Mass Production, 1800–1932: The Development of Manufacturing Technology in the United States*. Baltimore, MD: Johns Hopkins University Press, 1984.

Hughes, Agatha C., and Thomas P. Hughes, eds. *Systems, Experts, and Computers: The Systems Approach in Management and Engineering, World War II and After*. Cambridge, MA: MIT Press, 2000.

Hughes, Thomas P. *Networks of Power: Electrification in Western Society, 1880–1930*. Baltimore, MD: Johns Hopkins University Press, 1983.

Kesan, Jay P., and Rajiv C. Shah. "Fool Us Once, Shame on You—Fool Us Twice Shame on Us: What We Can Learn from the Privatization of the Internet Backbone Network and the Domain Name System." *Washington University Law Quarterly* (2001): 79.

Khambata, Adi J. *Introduction to Large-Scale Integration*. New York: Wiley-Interscience, 1969.

Kidder, Tracy. *The Soul of a New Machine*. Boston: Little, Brown, 1981.

Kidwell, Peggy A., and Paul E. Ceruzzi. *Landmarks in Digital Computing: A Smithsonian Pictorial History*. Washington, DC: Smithsonian Press, 1994.

Kilby, Jack S. "Invention of the Integrated Circuit." *IEEE Transactions on Electron Devices* 23 (1976): 648–654.

Kita, Chigusa. "J.C.R. Licklider's Vision for the IPTO." *IEEE Annals of the History of Computing* 25 (3) (2003): 62–77.

Knuth, Donald. *The Art of Computer Programming. Vol. 1: Fundamental Algorithms*. Reading, MA: Addison Wesley, 1973.

Krol, Ed. *The Whole Internet Users' Guide and Catalog*. Sebastopol, CA: O'Reilly & Associates, 1982.

Levy, Steven. *Hackers: Heroes of the Computer Revolution*. New York, 1984.

Licklider, J. C. R. "The Computer as a Communications Device." *Science and Technology*, April 1968.

Licklider, J. C. R. "Man-Computer Symbiosis." *IRE Transactions on Human Factors* 1 (March 1960): 4–11.

Liebowitz, Stan J., and Stephen E. Margolis. *Winners, Losers, and Microsoft: Competition and Antitrust in High Technology*. Oakland, CA: Independent Institute, 1999.

Lundstrom, David E. *A Few Good Men from UNIVAC*. Cambridge, MA: MIT Press, 1987.

Manes, Stephen, and Paul Andrews. *Gates: How Microsoft's Mogul Reinvented an Industry, and Made Himself the Richest Man in America*. New York: Doubleday, 1993.

Metcalfe, Robert M. "How Ethernet Was Invented." *IEEE Annals of the History of Computing* 16 (1994): 81–88.

Metropolis, N., J. Howlett, and Gian-Carlo Rota, eds. *A History of Computing in the Twentieth Century*. New York: Academic Press, 1980.

Mills, Mara. "Hearing Aids and the History of Electronics Miniaturization." *IEEE Annals of the History of Computing* 33 (2) (2011): 24–44.

Mims, Forrest, III. "The Tenth Anniversary of the Altair 8800." *Computers and Electronics* 58–62 (January 1985): 81–82.

Moore, Gordon E. "Cramming More Components onto Integrated Circuits." *Electronics*, April 19, 1965, 114–117.

Moore, Gordon E. "Microprocessors and Integrated Electronics Technology." *Proceedings of the IEEE* 64 (1976): 837–841.

Moore School of Electrical Engineering. University of Pennsylvania. *Theory and Techniques for Design of Electronic Digital Computers: Lectures Given at the Moore School of Electrical Engineering, July 8-August 31, 1946*. Philadelphia: University of Pennsylvania, 1947–1948; reprint: Cambridge, MA: MIT Press, 1985.

Murray, Charles. *The Supermen: The Story of Seymour Cray and the Technical Wizards behind the Supercomputer*. New York: Wiley, 1997.

Naur, Peter, and Brian Randell. *Software Engineering; Report on a Conference Sponsored by the NATO Science Committee*, October 7–11 1968. Garmisch, Germany: NATO, 1969.

Nelson, Theodor. *Computer Lib*. South Bend, IN: Ted Nelson, 1974.

Newman, Bruce. "Apple's Third Founder Refuses to Submit to Regrets." *Los Angeles Times*, June 9, 2010.

Norberg, Arthur L. High Technology Calculation in the Early Twentieth Century: Punched Card Machinery in Business and Government. *Technology and Culture* 31 (October 1990): 753–779.

Norberg, Arthur L., and Judy O'Neill. *Transforming Computer Technology: Information Processing for the Pentagon, 1962–1986*. Baltimore, MD: Johns Hopkins University Press, 1996.

November, Joseph. *Digitizing Life: The Rise of Biomedical Computing in the United States*. Baltimore, MD: Johns Hopkins University Press, 2011.

Noyce, Robert, and Marcian Hoff. "A History of Microprocessor Design at Intel." *IEEE Micro* 1 (February 1981): 8–22.

Ogan, Christine, and Randall A. Beam. Internet Challenges for Media Businesses. In *The Internet and American Business*, ed. William Aspray and Paul E. Ceruzzi. Cambridge, MA: MIT Press, 2008.

Pake, George. "Research at XEROX PARC: A Founder's Assessment." *IEEE Spectrum* (October 1985): 54–75.

Pearson, Jamie Parker, ed. *Digital at Work*. Bedford, MA: Digital Press, 1992.

Pugh, Emerson. *Building IBM: Shaping an Industry and Its Technology*. Cambridge, MA: MIT Press, 1995.

Pugh, Emerson W., Lyle R. Johnson, and John H. Palmer. *IBM's 360 and Early 370 Systems*. Cambridge, MA: MIT Press, 1991.

Randell, Brian, ed. *The Origins of Digital Computers: Selected Papers*, 2nd ed. Berlin: Springer-Verlag, 1975.

Randell, Brian. "The Colossus." In A History of Computing in the Twentieth Century, ed. N. Metropolis, J. Howlett, and Gian-Carlo Rota, 47–92. New York: Academic Press, 1980.

Redmond, Kent C., and Thomas M. Smith. *Project Whirlwind: The History of a Pioneer Computer*. Bedford, MA: Digital Press, 1980.

Reid, Constance. *Hilbert*. New York: Springer-Verlag, 1970.

Reid, T. R. *The Chip: How Two Americans Invented the Microchip and Launched a Revolution*. New York: Simon & Schuster, 1985.

Rifkin, Glenn, and George Harrar. *The Ultimate Entrepreneur: The Story of Ken Olsen and Digital Equipment Corporation*. Chicago: Contemporary Books, 1988.

Ritchie, D. M. "Unix Time-Sharing System: A Retrospective." *Bell System Technical Journal* 57 (1978): 1947–1969.

Roberts, H. Edward, and William Yates. "Exclusive! Altair 8800: the Most Powerful Minicomputer Project Ever Presented: Can Be Built for under $400." *Popular Electronics* (January 1975): 33–38.

Rojas, Raul, and Ulf Hashhagen, eds. *The First Computers: History and Architectures*. Cambridge, MA: MIT Press, 2000.

Rosen, Saul. "Electronic Computers: A Historical Survey." *Computing Surveys* 1 (March 1969): 7–36.

Salus, Peter. *A Quarter Century of UNIX*. Reading, MA: Addison-Wesley, 1994.

Sammet, Jean. *Programming Languages: History and Fundamentals*. Englewood Cliffs, NJ: Prentice Hall, 1969.

Schein, Edgar H. *DEC is Dead: Long Live DEC: The lasting Legacy of Digital Equipment Corporation*. San Francisco: Berrett-Koehler Publishers, 2003.

Shannon, Claude E. "A Symbolic Analysis of Relay and Switching Circuits." *Transactions of the American Institution of Electrical Engineers* 57 (1938): 713–723.

Siewiorek, Daniel P. C., Gordon Bell, and Allen Newell. *Computer Structures: Principles and Examples*. New York: McGraw-Hill, 1982.

Smith, Douglas, and Robert Alexander. *Fumbling the Future: How Xerox Invented, Then Ignored, the First Personal Computer*. New York: Morrow, 1988.

Snyder, Samuel S. "Influence of U.S. Cryptologic Organizations on the Digital Computer Industry." *Journal of Systems and Software* 1 (1979): 87–102.

Snyder, Samuel S. "Computer Advances Pioneered by Cryptologic Organizations." *Annals of the History of Computing* 2 (1980): 60–70.

Spicer, Dag. "Computer History Museum Report." *IEEE Annals of the History of Computing* 30 (3) (2008): 76–77.

Standage, Tom. *The Victorian Internet: The Remarkable Story of the Telegraph and the Nineteenth Century's On-Line Pioneers*. New York: Walker, 1998.

Stern, Nancy. *From ENIAC to UNIVAC: An Appraisal of the Eckert-Mauchly Computers*. Bedford, MA: Digital Press, 1981.

Swisher, Kara. *AOL.COM: How Steve Case Beat Bill Gates, Nailed the Netheads, and Made Millions in the War for the Web*. New York: ThreeRivers Press, 1998.

Torvalds, Linus, and David Diamond. *Just for Fun: The Story of an Accidental Revolutionary*. New York: Harper, 2001.

Turing, Alan. "On Computable Numbers, with an Application to the Entscheidungsproblem." *Proceedings of the London Mathematical Society*, Series 2, 42 (1936): 230–267.

Turner, Fred. *From Counterculture to Cyberculture: Stewart Brand, the Whole Earth Network, and the Rise of Digital Utopianism*. Chicago: University of Chicago Press, 2006.

U.S. Department of Commerce. National Telecommunications and information Administration. U.S. Principles on the Internet's Domain Name and Addressing System. http://www.ntia.doc.gov/other-publication/2005/us-principles-internets-domain-name-and-addressing-system.

U.S. National Security Agency. "The Start of the Digital Revolution: SDIGSALY: Secure Digital Voice Communications in World War II." Fort Meade, MD: NSA, n.d.

Veit, Stan. *Stan Veit's History of the Personal Computer*. Asheville, NC: World-Comm, 1993.

von Neumann, John. First Draft of a Report on the EDVAC. Philadelphia, PA: Moore School of Electrical Engineering, University of Pennsylvania, June 30, 1945.

Waldrop, M. Mitchell. *The Dream Machine: J.C.R. Licklider and the Revolution That Made Computing Personal*. New York: Viking Press, 2001.

Wang, An. *Lessons*. Reading, MA: Addison-Wesley, 1986.

Watson, Thomas, Jr. *Father, Son & Co.: My Life at IBM and Beyond*. New York: Bantam Books, 1990.

Weizenbaum, Joseph. *Computer Power and Human Reason*. San Francisco: Freeman, 1976.

Wexelblatt, Richard L., ed. *History of Programming Languages*. New York: Academic Press, 1981.

Wiener, Norbert. *Cybernetics, or Control and Communication in the Animal and the Machine*. Cambridge, MA: MIT Press, 1948.

Wilkes, Maurice V. "The Best Way to Design an Automatic Calculating Machine." In *Computer Design Development: Principal Papers*. ed. Earl Swartzlander, 266–270. Rochelle Park, NJ: 1976.

Wilkes, Maurice V. *Memoirs of a Computer Pioneer*. Cambridge, MA: MIT Press, 1985.

Williams, Bernard O. "Computing with Electricity, 1935–1945." Ph.D. diss., University of Kansas, 1984, University Microfilms 85137830.

Winograd, Terry. *Understanding Natural Language*. New York: Academic Press, 1972.

Wise, T. A. "IBM's $5,000,000,000 Gamble." *Fortune* (September 1966), pp. 118–123, 224, 226, 228.

Wolfe, Tom. "The Tinkerings of Robert Noyce." *Esquire* (December 1983): 346–374.

Wozniak, Steve. "Interview." *Byte* (January 1985): 167–180.

Yates, JoAnne. *Control through Communication: The Rise of System in American Management*. Baltimore, MD: Johns Hopkins University Press, 1989.

Zachary, G. Pascal. *Endless Frontier: Vannevar Bush, Engineer of the American Century*. New York: Free Press, 1997.

Zuse, Konrad. *The Computer—My Life*. New York: Springer-Verlag, 1993.

INDEX

PAUL E. CERUZZI is a Curator at the National Air and Space Museum, Smithsonian Institution, Washington, D.C. He is the author of *A History of Modern Computing*, *Internet Alley: High Technology in Tysons Corner, 1945-2005*, both published by the MIT Press, and other books.